*ENTRIES (MAXIMALISM)* documents the art that emerged in the late 1970's and early 1980's. Written as a chronicle, the entries in this volume record the observations of a major critic who unencumbered by theoretical bias, knowingly explores the new developments of a rapidly changing art world.

By skillfully meshing the nuanced meaning of recent art with privileged information concerning its makers, Robert Pincus-Witten's narrative constructs an intricate mosaic of today's controversial aesthetic endeavors. Sweeping aside such ephemeral cognomena as "new image," "neo-expressionism," "trans-avant-garde," "new figuration," etc., the term MAXIMALISM, might well become the definitive label of the 1980's.

Profiled in this book are Julian Schnabel, David Salle, Robert Longo, Gary Stephan, Eric Fischl, and Siah Amarjani, among others.

ROBERT PINCUS-WITTEN was the first critic to challenge the formalist hegemony of the 1960's by indicating that, as a movement, it had been superseded by a generation of brilliant artists. The support of these new figures led to a ten-year critical campaign at *Artforum* during its most pugnacious years, when at length he served as Senior Editor. Currently Associate Editor of *Arts Magazine,* he is a Professor of Art History at Queens College, City University of New York, and is on the faculty of The Graduate Center, CUNY. His *POSTMINIMALISM: American Art of the Decade* (Out of London Press, 1977), quickly became the standard reference on the works of Serra, Hesse, Sonnier, Benglis, Burton, et al.

ART AND ART CRITICISM SERIES EDITOR

Rita Di Pace

ROBERT PINCUS-WITTEN

# ENTRIES
# (MAXIMALISM)

## Art at the Turn of the Decade

OUT OF LONDON PRESS
New York • Norristown • Milano

Cover photograph by Charles H. Traub.

*Library of Congress Cataloging in Publication Data*

Pincus-Witten, Robert.
Entries (Maximalism): art at the turn of the decade.

1. Art, Modern—20th century. 2. Avant-garde (Aesthetics—History—20th century. I. Title.
N6493 1970.P56 1983 709'.73 82-62194
ISBN 0-915570-20-3


Published by Out of London Press, Inc., New York, Norristown & Milano.

First printing 1983.
Printed in the United States of America.

For my mother

CONTENTS

# ILLUSTRATIONS

37. Lucio Pozzi, 29 Models Descending an Art School Staircase, 1971.
38. Lucio Pozzi, The Migration, 1981.
39. Richard Serra, Different and Different Again, 1973.
40. Richard Serra, Shift, 1970-72.
41. Richard Serra, T. W. U., 1980.
42. Jackie Winsor, Small Circle, 1969.
43. Alberto Giacometti, Suspended Ball, 1930-31.
44. Eva Hesse, Vertiginous Detour, 1966.
45. Jackie Ferrara, The Ball and Saucer, 1972.
46. Jackie Winsor, Nail Piece, 1970.
47. Jackie Winsor, Bound Grid, 1971-72.
48. Jackie Winsor, Bound Logs, 1972-73.
49. Michael Hurson, Thurman Buzzard's Apartment, 1973-74.
50. Michael Hurson, Edward and Otto Pfaff, 1974-75.

# INTRODUCTION

The criticism said to matter in our time skims style from theory, skin from bones. The model, no matter how often denied, is Clement Greenberg. It was he who first promoted a set of conditions for painting or sculpture predicated on values exclusive to painting and sculpture. He saw that one defined the other as separate but equal entities. What painting was, sculpture was not; what sculpture was, painting was not; and taken together, all that painting and sculpture were, was not anything else — neither literature, nor politics, nor psychology; nor did they play themselves out in time.

The honing of these perceptions — those interminable discussions about planarity, frontality, edge tension, and on and on — came to be what we call Greenberg's "modernism," a machinery, at length, ratified by a selective reading of the history of European and American art from Manet on. But Greenberg himself was indifferent to certain conventional methods of art history, especially that area of speculation connected to the non-formal meaning of images, iconography.

While, for sure, perplexing contradictions remain — as when, say, sculpture aspires to a pictorial condition (the work of David Smith) or when he talks about "quality" (a kind of inescapable result when all components are in proper mix) — still, Greenberg's great achievement was not the designation of a diagnostic profile of modernist desirables, but in his having pointed to the artists who really mattered to his generation. His methodology conferred on his perceptions an acute percipience. Though he would not say it this way, Greenberg, the formalist, really mattered because he pointed to shifts in style and these recognitions had the effect of even further intensifying such dislocations.

By the late '70s, it was thought that the time for the application of Greenberg's methodology had passed, that successive waves of second and third generation formalist art simply failed to obtain, failed to be interesting. But it was the painting and sculpture themselves that were uninteresting, not the critical theory which believably corresponded to the values of the first generation. When latter day, formalist art was finally seen to be the academic thing that it was, we were at a moment when all the other stuff being made at that time came to be interesting, if only through negative definition. Whatever it was, and it was many things, at least it was said not to be formalist art. But that it is the Pluralism it is said to be is also in doubt. Still, what this art (whatever it is) lacks is the critical muscle tone that Greenberg's formalist theories gave to the first generation.

There is a ten year interregnum, 1966-76, between the disintegration of a formalist art and the art of our own moment. I tried to account for this episode in *Postminimalism.* Here is postulated the possibility of an osmotic interface between a collapsed form of painting and sculpture deemed "pictorial/sculptural"; the emergence of an abstract, information-based epistemology; its ontological counterpoises, body art and the conceptual theater. That these three modes roughly emerged in so chronological a sequence is seriously doubtful. Though of one thing we can be sure: when one wants to find something, when one knows what one is looking for, for sure one will find it.

It would be nice were one able to postulate such a gremlin-free mechanism at work in the years from the middle of the '70s to the present moment. Yet one sees that this is hardly the case. Among the reasons militating against so positivist un-nuanced a reading is that the ingathering, embracing character of Postminimalism led to an even more open vanguardism — so much so as, in the end, to defy categorization.

In the absence, then, of a credible theoretical correlative to the art following the mid-'70s, an art shruggingly identified as pluralistic, baffled criticism fell back upon the very methods which Greenberg proscribed and circumvented — especially iconographic analysis, though Wölfflin-derived formalist analysis had never really been dislodged as a component of the task.

With Pluralism then, if it is Pluralism, many novel features emerged: a refreshed emphasis on Wölfflin; a stress on iconography; the emergence of a neo-formalism based on structuralism and linguistic master-texts; and finally, the chronicle (to which I attribute, at this moment, a value quite beyond the reaches of quotidian journalism). That the chronicle will maintain a heightened value beyond this moment is doubtful — something in our culture today has lent a certain weight to a journalistic exercise.

It is not that I am out of sympathy with the desire to formulate a linguistic theory of contemporary art. The bankruptcy of a specious and sentimental romance of art and its concomitant maintenance of the sanctimonious status of artists, as if they, unlike anyone else, are above discussion, is a salutary demythification. That these desentimentalizations are desireable is unquestioned; to assert that they would occur anyway as a function of an inescapable force of history — some kind of manifest destiny against which free will is powerless — strikes me as specious, and in the end arbitrarily determined.

Also, there is my distaste for the formalist pose itself that makes it seem as if there is a specific content to art, a final and absolutely correct reading. I do not see how this can be since, after all, we are dealing with the imperfect world of experience as well as with the fact that we really do not know what art is; and, in the absence of such knowledge, how can we ever speak of a proper reading for it?

Last, all this is undermined by a simple pragmatic fact — critics are not as good as their methods, no matter how refined these methods are. Their worth, instead, is measured by the art they choose to defend, and whether that defense is made early enough. Greenberg's gift did not consist of the invention of a yeasty set of modernist desirables — nor was it a function of simply writing well. Greenberg's achievement resided squarely in the fact that while thinking well and writing well, he was pointing to the significant artists. And if you don't do the last thing best, the first two just don't matter.

I see the critical task as being essentially that of pointing to the new, and if these gesticulations thrust me into the position of having to call upon dubious art historical methodologies, well then so be it. I prefer to call up those time-honored modes aligned with the freshness of chronicled experience. The real issue at hand is not new modes of criticism, but what happened to painting and sculpture during the last few years. Sculpture had the edge on painting in the late '60s owing to David Smith's early use of a pictorial vehicle for sculptural expression. By the end of the '60s, pure painting was hobbled by its bondage to formalist values insofar as it irresiliently carried the values of first generation Abstract Expressionist painting through a second and third phase. As this shrivelling occurred, so did sculpture flourish. Concurrently, all the unsatisfied need to experiment was cut away from the easel enterprise and was manifested as some kind of sculptural polymorph.

Although less liable of conventionalization, the appeal of a pictorialized sculpture began to fail of late, and sculptors sought a more intense sculpture by reverting to the latency of Constructivism, Constructionism. An unapologetic construction of rigorously architectural character received particular emphasis through the end of the '70s. The Constructionists — Hall, Aycock, Ferrara, Acconci, Oppenheim, Armajani, et al. — did much to preserve the syntax of Minimalism. Apart from the feeling for the elemental character of the substances they used (which would relate them to figures like Serra and Andre), their work also went far to foster a taste for architecture-as-art.

This largely happened for two reasons: the blueprints or working diagrams for Constructionist sculpture demonstrated a high degree of preexecutive awareness and, as such, sought the diagram, the hand-drawn blueprint, as a central expression of the style. This condition is also true of Minimalist arts generally, with the following warning: the Minimalist blueprint is essentially scaleless and placeless — it is a purely refrigerated abstraction ratified in the mind as a Platonic ideal. By contrast, the Constructionist diagram or blueprint is specific as to scale, specific as to site, and generally invokes an impure pictorial metaphor rather than the pure geometry of Minimalism. The aberrant element in all of this is the metaphorical mutant, since what it does is to value those literary and psychological qualities presumably expunged from Formalism. Thus, the metaphor of the Constructionists allowed for a revived interest in architectural drawing at the turn of the decade, and architects-as-artists benefited from this skewing of esteem (from sculpture to their work), provoking an essentially architectural polemic focused on what has come to be called Post-Modernism.

All this, and more, is confounded with several colliding streams that originated in quite different spheres of activity. Where Constructionism comes from is easily grasped; architectural style less so. Simply, the latter is reacting to the rejection of applied ornament to architectural problems germane to the International Style Between the Two Wars. Instead of viewing ornament and applied design as taboo, recent architects began to see ornament as a manneristic means of generating forms that answered functional requirements. Thus began a celebratory and eclectic pillage of the history of art for design-cum-functional motifs by a group of architects reacting against the crisp and clean machine forms of a precisionist International Style, a style loosely referred to as "modern"; hence these architects were called "post-modern."

Their eclecticism, on one hand, coincided with the personalistic metaphor of the Constructionist sculptors, fusing these two strands. On the other hand, these snowballing options all feed (as they are being fed) the neo-Primitivistic Expressionist positions in painting which contemporaneously emerged at the turn of the decade, deeply stirred by recent developments in German and Italian painting.

A rough outline of the art that emerged between the American Bicentennial and the turn of the decade continues forward from the three elements of Postminimalism, to the advent of Constructionism, Architecture-as-Art, and finally, neo-Primitivism and Expressionism. All this latterly development I am calling Maximalism. With these developments, the reign of sculpture, dominant in the '60s and '70s, appears now to have been superceded by the successful reemergence of a new Primitivistic painting as the fundamental vehicle of expression of the present, with all that that signifies as regards the appurtenances of painting such as the value of facture, the return to allegory, etc.

The democratic vision of contemporary history, still colored by the ideals of John Dewey, has made a case for egalitarian Pluralism out of all of this. Such a view seems ill-advised since the character of all of these recent modes is an intense awareness of the history of modern art. The essential premise of Pluralism is the separateness of the art experience from the desire to make art imbued with a sense of historical continuity. But, in the absence of an art made in a state of historical awareness, the art experience is no more than mere neurological stimulation. Indeed, Dewey's term "art experience" is a contradiction since "art" implies historical awareness while "experience" rejects it. In short, "art experience" does not exist outside of historical awareness — hence criticism and history — whereas experience may exist free of such attachment.

For the Pluralist, the essential antagonist is nature; for the contemporary artist, the essential antagonist and malleable form of art is the very awareness of art history itself, an awareness most often encountered as an irony.

One of the things that goes unnoticed in the so-called Pluralist position of the present day is just how much of it is merely a Philistine backlash against Postminimalism. In terms of many painters, though not all, this has sanctioned an academic figure painting with a concomitant reliance on the illusionism derived from the Renaissance and the Baroque — the space box of the one, the chiaroscuro and tactile appeals of the other, and the emphasis on figure composition of both. Alone, these issues remain moot; but allied to an ironic awareness of the history of contemporary art, ignition occurs.

Not that all contemporary figure painting is bad; indeed several of its most positive properties have contributed tellingly to the current mood. Figure painting of interest today has stressed the private talisman, a sign of an intensely peculiar personal episode. Perhaps the success of feminist art occasioned this. In so doing, the stress on the private talisman has been allied to the revived Abstract Expressionist focus traceable to the sculptural/pictorial collapse typical of early Postminimalism. Thus, the pictorial properties of Abstract Expressionism — space defined through the gestural allover or blurred optical fields — assert themselves over the academically derived illusionism of Renaissance and Baroque art. Still, the sources for this possibility synthesize private impulse and psychology with such formal properties, resulting in a neo-Primitivism, a new Expressionism which is an ironic commentary on the history of art, though nonetheless masterful for all that.

A paradox: The confines of Maximalism are smaller and tighter than those of Postminimalism. Under the latter's sway, a loss of species occurred — no more painting defined conventionally, no more sculpture defined conventionally. Instead, a broad spectrum of open type was invented in these species — earthwork, say, as sculpture; arithmetic and grammar substituted for painting; performance or story art for either — a vast berth the understructure of which was set in language.

But now, with all that past and under a Maximalist banner, we find that painting and sculpture, for all that seems gratuitious and thrown-in with the kitchen sink, is once more a closed species returned home to normal typology. Thus, Maximalism, despite its brave name, is perhaps a more conservative artistic consciousness — let's say it, style — than its predecessor.

Style, for me, is the ultimate residue of art; it is the intellectual construct, the fiction, that allows for, accounts for, infra-and-international transposability. All this points to the fact, however shocking it is to read, that what one ultimately seeks is "signatures" — talisman/objects of the art-experience. But discussion of the experience details self important lists of enthusiastic response; or, argues for a transcendental defense of art, that is "art-as-religion," something one either believes or does not. It is a matter of faith. I do not say I lack this faith. I only say it is futile to pretend that articles of faith can rationally be discussed in other than an historical way, as a branch of the history of religion. All transcendental defense, like religion, ends up with the belligerent "my book is better than your book."

I am aware of the rationalized character of this outline. Its real justification is far more skeptical and fatalistic than those indicated in a specious mechanic of stylistic evolution. Ultimately, we can never know that what one ever says about art is true. So, if sheerly factual writing is doubtful from the outset, then for sure speculative histories are even more suspect. My skepticism, then, must reject the formulation of a closed theory of history, including my own.

With this in view, it is gratuitous to broadcast a new "ism" of the kind found in Postminimalism, Post-Modernism, Pluralism, or even Maximalism. If a so-called archaeology of contemporary art, the excavation of straight fact, is not apt to be true, then how much more false would a theoretical mechanism be?

But one mode of discourse in the arts is least liable of untruth — the chronicle. In it, one is really recording what one is thinking or feeling, though the facts may be wrong and the feelings unwarranted. To say all this, to note that they may possibly be wrong is no demonstration that they are wrong — nor, for that matter, right. Statements of right or wrong function outside both the history of art and criticism. Morally, they appertain to law; as abstraction they are a function of logic.

Thus, I view the chronicle — the record least liable of incorrectness set into a cultural matrix of absolute ambiguity — as a literary effort convergent to the artwork of the turn of the decade. In *Postminimalism*, the reprinting of essays sought to capture the discourse of the artists in question; the republication of the essays that form *Entries (Maximalism)* is spurred by the parallel status of these chronicles to the art which provoked them.* While not being art, they are perhaps not artless.

*Entries: Palimpsest and Pentimenti, *Arts Magazine,* June 1980; Entries: Maximalism, *Arts Magazine,* February 1981; Entries: Syles of Artists and Critics, *Arts Magazine,* November 1979; Entries: Two Voices, *Arts Magazine,* April, 1979; Entries: Sheer Grunge, *Arts Magazine,* May 1981; Julian Schnabel: Blind Faith, *Arts Magazine,* February 1982; David Salle: Holiday Glassware, *Arts Magazine,* April 1982; Defenestrations: Robert Longo and Ross Bleckner, *Arts Magazine,* November 1982; Gary Stephan: The Brief Against Matisse, *Arts Magazine,* March 1982; Eric Fischl: Freud Lives, *Arts Magazine,* September 1981; Strategies Worth Pondering: Bochner, Shapiro, LeWitt, *Arts Magazine,* April 1978; Camp Meetin': The Furniture Pieces of Scott Burton, *Arts Magazine,* September 1978; Benni Efrat: The White Rectangle, *Arts Magazine,* April 1977; Siah Armajani: Populist Mechanics, *Arts Magazine,* October 1978; Lucio Pozzi: The Continence of Lucio, *Arts Magazine,* March 1980; Entries: Oedipus Reconciled, *Arts Magazine,* November 1980; Winsor Knots: The Sculpture of Jackie Winsor, *Arts Magazine,* June 1977; Michael Hurson: A Fabric of Affinities, *Arts Magazine,* June 1976.

# ENTRIES:
# PALIMPSEST AND PENTIMENTI

*7 February 1980*

Let me try to be more precise as to what I mean by style. Style represents a moment of freeze and shrinkage, an instant when artistic intuitions are perceived as possessing a consistent diagnostic profile. Indeed, it is the overlapping, repetitious character of such diagnostic elements that makes it seem that there is, after all, such a thing as a style in the first place. To be interested in this notion of style as generalized phenomenon in no way precludes my greater interest in "pre-style," that is, those moments and expressions of artistic intuition risked even prior to shrinkage and freeze.

*23 March 1980*

A certain kind of problem continues to nag at what we tend to consider by now a provincial preoccupation — the problem of "quality," in the straight Greenberg use of the word. Astonishing as it may be, many people still haven't gotten around to admitting that "quality" most often means the attribution of a specious transcendency to work that merely corresponds to an *a priori* set of features concerned wholly with formal components. More, such attributions are probably no more than a nostalgic sentimentality inescapable to dying bourgeois aestheticism.

The beginning of the Eighties is marked by a rejection of the open modes that had been sponsored and favored throughout the Seventies. As always, when this occurs, smug status-quoers (who never flirted with, let alone abandoned, unstable modes) act either as if they had been right all along not to have done so; or, even more gratingly, as if the open modes just never had happened in the first place.

The smug entrenched formalist ethos of our Eastern seaboard is a function of the history of more than two and a half centuries of painting here. Thus, Smibert and Blackburn and Copley form a root layer of loam nourishing formalist prejudices favoring painting as a stable and closed species. The historical couche of the West Coast, behavioral and performance oriented, is scarcely twenty years old. That means the local formalist bias is rooted far less securely. But where will it lead? Even on the Western shore today, a fatigue with performance and open modes is sensed.

*9 April 1980*

Put through one's paces by the seventeenth century — chiaroscuro graphics. The most common gift among artists is a graphic one; the rarest, that of color. After three centuries and more of inculcation, the young art student's intuitions drift toward dark-light contrasts, especially of the kind derived from Baroque chiaroscuro and the chiaroscuro print. An updated version leads directly to the steel plate engraving of Washington's portrait on the dollar bill. Here, the linear swellings and undulations attempt in a tactile way to follow the modulations of surface of the thing depicted as it exists in real life, through denoting the processions and recessions of the surface of the model in nature. Even when the young artist is drawn to making abstract images, such kinds of modulations argue for what Greenberg correctly saw was a lost figuration, a "homeless representation." Yet students know that painting means color. So, going against their most natural graphic gifts, they assign variant reds and yellows and blues to the chiaroscuro modulations. Such allocations are most arbitrary, even unnatural. The most natural expression of graphics is line; of painting, color. Line is to graphics as color is to painting. Chiaroscuro, a function of spatial illusion, is the enemy of both line and color as it is more naturally a tactile value and thus a function of sculpture.

The expressionist representationalism of student artists most often results from a desire to accommodate a seventeenth-century graphic model (most often Rembrandt) to an Impressionist or sensationist-derived color. The upshot is that this kind of representationalism works the turf of Lovis Corinth, itself a reworking of Max Liebermann — German artists who tried to escape an ingrained cultural thrall to Rembrandt through tying it to Impressionist notions of color.

The pure untrammelled expression of line is equal to that of color; chiaroscuro, a tactile value, always gets in the way. Matisse above all saw this most clearly. Daunted by the success of Cubism, Matisse attempted to modify his clear grasp of color, to force it to conform to a new analytical system, that of Cubism — a graphics-derived style. As a result he quite inescapably made "failed paintings." But, for Modernism, "failed paintings" are our greatest expressions, precisely because they engage us *à rebours.*

*6 April 1980*

A reception to mark the opening of a group show of "Boonies" —Ross Bleckner, Julian Schnabel, and David Salle. The latter's work somewhat resembles that of Eric Fischl (Plate 1) — they were former California schoolmates — whose show opened that day across West Broadway at Edward Thorp's gallery. One could easily add to this list the name of Gary Bower, now with Max Protetch (indeed, Max is showing the important trends: in sculpture, the architectural constructionists; in painting, the contradictory evidence of the figurative disjunctionists). Schnabel is perhaps the strongest painter of the new stylistic coalition and he tells me about his imagistic snatchings — spolia, as it were — from the films of Jean Vigo —*Atalante, Zéro de Conduite*. Fragments of images from these films supply Schnabel with subject matter.

The key to the work of the "Boonies" is something that escaped my notice recently. And that is the problem (indeed, for them all) of the overlap, the superimposition, the palimpsest (Plate 2). These transparencies and superimpositions allow for a figurative painting denying narrative all the while. In this way, the essential weakness of representational painting is eluded, namely its inescapably illustrative properties. As always, when one thinks about it, models suggest themselves — among them the expressive "Transparencies" of Picabia painted after 1926 and, more magically, the pile-up of image on image found in cave painting.

I am hounded by a concern that notions of quality in art are merely nostalgic and sentimental hangovers — a kind of grief — associated

with a perhaps moribund, certainly threatened bourgeois culture.
I worry about the mesh of culture in the art of the Eighties, that sieving warp and woof. Can it be that all the "Boonie" painting, all that palimpsest and pentimenti, New Figurative expressionism still represents only the drag of old ideals (which in part it surely does)?
But perhaps rising in this couche is a sprinkling of young artists who do point the way. They most likely deal with photography or photographic notions in their work though they are not necessarily photographers. I suspect that in some way they are photographic collagists though the collaging may occur in an environmental way. Their work derives meaning from the fine-grinding Marxist machinery of Walter Benjamin's *The Work of Art in the Age of Mechanical Reproduction,* intuitions, it would seem, refurbished in the phenomenology of Adorno and the Frankfurt School and in French Structuralism, especially Foucault.
Of course, what I am alluding to is, yet again, a rigid theory of pure Formalist inevitability. Still, whether one likes it or not (and I hate it), there is the fascination of such a possibility, though one balks at all the tailoring and pruning necessary to demonstrate the validity of such a principle. But this idea does render impertinent many simple, received ideas about style, sentiment, personalistic iconography, the capitalist myth of a great hero, the morphological-similarity of Wöfflin as demonstrating content-similarity, iconography as a literary discipline, etc. The notion that we are just flotsam and jetsam of larger controlling forces of destiny like the one I drew renders all the above and more just so much sentimental rubbish, to be finally relegated to the dust-bin of history.
On one hand, all this then might be viewed as an updated Greenberg-like formalism. On the other hand, such an idea supplants morphology and iconography as the conventional methods of stylistic and/or art-historical method. What I'm talking about then may be a post-Marxist, post-phenomenological, in short, post-Structuralist methodology — as others claim it is — just out there in the world whether one likes it or not, supplanting willy-nilly through its sheer essential force these earlier attitudes and methods.

I would like to be able to believe the above easily, but can't. Mostly I've just recast the general drift of the arguments put forth by the critical "heavies" today. What they won't see (won't let themselves see) is that ideas are as subject to style as forms are. So all this post-modernist inevitability is as locked into bourgeois history as anything else — despite the delicious self-delusion that post-modernism eludes bourgeois history.

*24 April 1980*

Early modernism is an argument essentially conducted between representationalism and abstraction. Indeed, early modernism is itself the exact embodiment of this discourse with, ultimately, the gaining of the upper hand by an abstraction that had placed representationalism *en jeu* in the first place. What is most rarely noted in this idea of early modern is that, in answering the conditions imposed through the inevitable application of abstraction upon representationalism, form destroys itself, self-destructs. Take a simple example — Analytical Cubism. As a style it is defined, at least partly, as a geometrical analysis of natural form. Fernande's forehead then may be rendered (as indeed Picasso rendered it in 1909) as four rectangular planes. Yet in honoring the principle of analysis in Analytical Cubism, these planes might further fracture into eight facets and these, in turn, into sixteen and so on, so that in the end these fractures and facets — if honored as structural principle alone and then carried to a logical conclusion — would ultimately destroy an essential term of the argument, namely its representational quotient. Hence early modernist discourse was made only through partial discussion, through reticences and hesitations that, by and large, maintained the representational term placed *en jeu* by abstraction. With a few exceptions - Malevich, El Lissitzky, Rodchenko, perhaps Mondrian (begrudgingly) — do we find the representational term truly or fully capitulated to the sheer press, to the *force majeure* of abstraction.
Late modernism does not even admit the terms representational versus abstract as being pertinent any longer. They both are indicted terms. Our argument is said to be taking place somewhere else out there. But where? And how?

Modern method is born of a need to reify speculation. Such speculation is reified, is transformed into "real" observation and perception, that is, into "things" liable of discussion. Modern historical method then becomes one in which the following question is posed: What are the models for this *reification?* Hence, modern method — as distinct from those applied to or in the culture of earlier epochs (they were generally formal or iconographic) — is fundamentally linked to criticism, critical observations, and critical speculation.

ENTRIES:

MAXIMALISM

*17 September 1980*

Paris-New York: the failures of both cities are so manifest that they appear almost to mirror one another. Paris works insofar as the city is still respected: its public spaces are, to all intents and purposes, little abused while ours are, by contrast, in frightful disarray. We are witness to the destruction of our parks, our public transportation systems, our libraries, our very walls; the air we breathe is polluted by chemical fumes and tapedeck soundspill.

All that is absent, in certain measure, at least, from Paris. But there, there is only a sense of looming hopelessness. To earn one's daily bread, even for the bourgeoisie, simply means to tread water; and to do that is more than even the French have come to expect as "raisonnable." All around, the city already aged and decayed, ages and decays even more, providing yet a new burden — the baleful weight of its own history and urban mores. By moments, things get patched up, refurbished — changed *quartiers* and the like; but such changes, while they preserve the city a bit longer, also create the new preserves of the rich.

Bohemia positive and productive, creative Bohemia rather than picturesque poverty, that kind of Bohemia has ceased to exist in Paris. Instead, one gets urban Bohemia, a new type where there is only changelessness, no originality, and a marked absence of choice typical of an urban peasantry.

*17 October 1980*

Ours is so utterly a cinematographic culture that the pleasure afforded by a painting is, at best, that given when it is regarded as a kind of still photograph. Beyond this, this "still" is just another frame in the larger film sequence. Can this intuition connect with the curiously frozen-frame character of much New Wave painting — Robert Longo, say? Ironically the term "New Wave" has itself an additional cinematographic association insofar as it was first applied not to the contemporary punk style at our turn of the decade but to the young European film-makers of the post-Naturalist moment, Godard especially, in the late 1950s.

*25 October 1980*

Dinner last Monday night with Alain and Ariane Kirili. Touched by their open, even pro-American awareness, especially in the face of the prevailing anti-Americanism in France today, part of a continuing and unexamined pro-Soviet sentiment there. This fiber of modern French consciousness coupled with the disastrous academization of School of Paris values goes far in explaining the wretchedness of contemporary French art, an abasement that not even the growing collection of good post-World-War-II art at the Beaubourg is liable to correct — certainly not now that it will return to the direction of a French government functionary.

Gained some insight into Kirili's sculpture. He is a "pure" ironsmith — forge-beaten iron. In our elevation of David Smith and company to the rank of the immortals we had forgotten that the oxyacetylene-welded joint and the anvil-forged passage are really two separate strains of the Constructivist idiom. I am tempted to write Constructivist "ethos" insofar as Kirili's Constructivism has "ethically" chosen to remain true to the earlier and purer Constructivism of forged iron and the artisanal traditions of "la ferronnerie," as the French call forged ironwork.

Kirili's works appear to be the first attempts at forged iron sculpture (thought of as autonomous sculpture) since the providential moment of Gargallo, Gonzales, and Picasso. Gargallo was technically progressive (forged iron) but iconographically reactionary (Synthetic Cubist commedia dell'arte). Gonzales and Picasso were technically and iconographically progressive (forged iron plus Surrealist metamorphosis). And what does one make of the rare beaten iron pieces by Brancusi also dating to c. 1930? These also are technically and iconographically progressive (forged iron plus the vestigial stickfigure).

But we must try to remember (since we appear to have forgotten) that the Blacksmith and David Smith represent two distinct strains, perhaps even antithetical ones, of the history of Constructivism, though always presented as continuous.
Kirili works partly in Philadelphia with the son of the great forged-iron craftsman Samuel Yellin as well as with another craftsman of Russian origin on the outskirts of the city. Yellin, hardly a household name, still is remembered as the maker of the grand pseudo-Renaissance lamps that used to hang on the Medician walls of the great bank at 73rd and Broadway. (Where are they now and who is responsible for that vandalism?)
The oddly "slipped and disjunctive" iron bases for Kirili's totemic iron sculptures force the viewer to circle about these otherwise lean and rigorously frontal shafts. And, more recently, Kirili's beaten iron elements curl and double back upon one another's shafts, forming an oddly graceful female icon despite the austere, "fruste" character of Kirili's poundings (Plate 3).

*1 November 1980*

Returning from quick jaunt to Cranbrook . . . . But, first a run-up to Lansing to see Michael Heizer's trivial, if orotund, effort made out of hyperscale Cor-ten concerning the epistemic examination of whole disks and their sub-divisions.
The whole thing very G.S.A. service plaza and set up in front of the pseudo-baronial capitol buildings. Then back to the lawns of Cranbrook to see Dennis Oppenheim's sprawling metaphorical and expressionist piece about mining — more exactly about ore alchemy. The work is unrepentant (and properly so) about its fusion of the mecanomorphs of the *Large Glass* to the open construction of the *Monument to the Third International.*
Then I just started talking to the students, picking up a conversation begun about two years ago when we were all concerned with the death of Constructivism, now run full cycle to the resuscitation of Constructivism. Several strains may account for the raising of this mighty Lazarus. One was the mode of Constructivist self-preservation that I called "constructionism" — real sculpture in the scale, methodology, and iconography of real architecture.

Another emerges as part of the general revival of metaphor and allegory so marked in our moment of painting as the prime vehicle of cultural expression. How different this is from the '60s and '70s when sculpture was its prime vehicle.
Sculpture today is inflected by the vaguely programmatic and episodic adventures of pictorial structures shot through with complex influences and private analogies. Oppenheim, say, and Alice Aycock and Vito Acconci are representative of the trend. The new models, then, for Construtivist sculpture have become Duchamp and *The Green Box* — the mecanomorphic adventures of the elements of *The Large Glass* as clues for the part-to-part adventures of the new allegorical Constructivism.
A return to metaphor and allegory perforce are painterly considerations in that they have always lain in the purview of the constituent features that defined the species of painting itself. This return strikes one as especially vivid, considering the long Minimalist *emprise,* a mode dedicated to an expression that functioned within the prerogatives of sculpture, no matter the media in which the style was embodied.
In the 1970s painting and sculpture were deeply marked by the crafts spillover. As Michael Hall acutely noted, had pattern painters begun to paint but a year or two previous to when they began, they would really have chosen to be potters or weavers; but fortuitous coincidence allowed them to renounce crafts — well, they never really considered crafts — for arts. Typically, this 70s spillover took place in California, a turf far less liable to be finicky about species-spill at the edge.
The night before, Eleanor Antin's continuing and touching musings of the aged Antinova — the great black face in the imaginary snow bank of the Imperial Ballet in exile. Antinova recollecting her life with Diaghilev, a performance installation including photographs in the manner of Baron de Meyer, in Antinova's great first mime role as *L'Esclave* (Plate 4) — and thence her suite of financial disasters but *succès d'estime* for the Ballets Russes de Monte Carlo of the immediately post-World War-I seasons — Pocahontas, Prisoner of Persia, The Hebrews, Before the Revolution. Her account culminates in the telling of the events surrounding the débâcle of her

magisterial *Before the Revolution* (New York City revived it in 1979). In this effort Antinova, the great black ballerina, played, of course, Marie Antoinette.

Stills of this role and others of her post-*L'Esclave* period were shown in a photographic mode owing all to Arnold Genthe whose photographic sense of "plastique" mediates Isadora and the American academicians Mario Korbel, Isidore Konti, and C. P. Jennewein.

The exhibition included page reproductions of her great volume of memoirs that typographically reconstitutes, as it were, the pages of the garnet buckram bound and colophonically syncopated volume of Karsavina's memoirs of a life spent in *Theatre Street*.

How oddly art works. On seeing Antinova's grand reconstruction of *Before the Revolution* last season in New York (it looms greater in memory than perhaps it really was, as does all memory of conceptual performance), I gave Antin/Antinova my copy of *Theatre Street* as a kind of souvenir of the event, writing a florid dedication signed "Graf Bobbi" (to invoke the idiotic adventures of the foppish counts Nicki and Bobbi, of the numerous Graf Bobbi and Graf Nicki *Witze* that circulated in the waning light of the Austro-Hungarian Empire). My chancery hand being what it is, led Antin/Antinova to read Graf as "Gray" and an odd new character emerged in her mind — a crippled Scot, Gray Bobby, a sweater manufacturer and protector of ballerinas. This character, as her story develops, is an easy touch — despite his Scottish roots —from whom Antinova's lovers and fellow dancers extirpated hundreds of bright woolens during, say, the '22 season. It's all wrong, of course, but in its way, absolutely right.

At the end of the evening, the aged Antinova, halting and misty-eyed with her self-glorifing vapors, acutely nudged her public at this, her final New York seance, in the direction of the autographed copies of the memoirs available at the exit, noting that she, this doddering and grandly magnificent wreck, is free for recitations, *thés* and *soirées*. The captivating ensnarer who bedeviled Diaghilev is still there.

Can I be the only one who is delighted by this admittedly collegiate nonsense so cleverly studded with barbs and intramural art-world

digs at the solemnities of East Coast Late Formalist dynamics turned by a pioneer West Coast performance artist? I expect so, as the audience for this kind of thing has to be a bit special, forearmed (as it were) with a certain awareness of the transient and pathetic histories of those staggering geniuses that once did really gravitate around Diaghilev.
As always, Antin writes and acts better than she ought in these circumstances, so that in the degree that she's so good but not good enough (in terms of the demands of the conventional theatre) the performance skirts failure. Which is part of its even greater beauty, it is so *divinement raté*. On taking leave of this grand disastrous wreck of a dancer, one could only grasp those gnarled, but still expressive, hands in one's own, raise them to one's lips, bow one's head, and murmur "Toujours la ballerina assolatissima."

*12 November 1980*

Helene Winer (ex-Artists Space) and Janelle Rearing (ex-Leo Castelli) open Metro Pictures on Mercer just below Houston . . . the space originally an afterhours disco club — very *Saturday Night Fever* — for garage mechanics — so, in a certain way, Metro Pictures is a bar with a gallery for a back room.
The evening essentially a happy but disquieting one: it definitely marks the death of the '60s. Henceforth, we of the '68-'72 set, no matter our good will, are of another, older generation. In the juke box light of the dance floor lit from below I could feel my laughing crow's feet deepen into wrinkles. This is no *plus ça change* moment but a different era, even if it's little history compared to big history. As 1922-25 was to 1912-14, so are the '80s when compared to '68-'72. But how can one say this without appearing condescending?
I'm repeatedly being fingered to pull together an exhibition of all the truly innovative artists including the young Italians from all the stables. If I do it, I'll call it Maximalism to make as strong a differentiation from the indurated and academized sensibility Minimalism of the '70s as possible.

*27 November 1980*

Anina Nosei, Helene Winer, Janelle Rearing, Mary Boone, Robert Miller, Angela Westwater, Max Protetch, Holly Solomon, Miani Johnson, Edward Thorp, Hal Bromm — the young dealers sometimes seem so much more interesting than the art they sell. The present scene is much changed, yet fully crystallized. The Italians quite different from the Americans — the former with their intense and acute sense of modern art history; the latter largely indifferent of this history but not necessarily ignorant of it. Rather, the Americans connect to the all-star mentors of a preceding generation. The young New Yorkers defiantly continue to admire Oldenburg and Warhol; the young Californians William Wegman, though he is scarcely older than they; and for them too, Warhol still casts a long shadow.

The current American mode "hypostatizes," "icon-icizes," "freezes" fleeting memory images of real people in real actions in real settings; in short, much of the American mode derives from the traditions of performance art, that is, the eccentric marginal art of the 1970s — a decade preeminently Sensibility Minimalist in flavor.

The New Wave Americans utterly reject the quietistic monasticism of Minimalist Sensibility painting (though they are, for all that, painters) and link up their work with transient glimpses of artists caught in performance activity during the '70s. But we are in the '80s, after all, and the '80s will be the decade of New Figuration, certainly the first few years of the decade and that's what counts. As the late '60s define the early '70s, so do the late '70s determine the early '80s.

## ENTRIES:
## STYLES OF ARTISTS AND CRITICS

*29 April 1979*

. . . Related issues: a student's inability or reluctance to read serious criticism is partly a reaction to a professoriate that lays — or laid (as I wonder how many professors really read it anymore) — great store by such writing. Is there a connection between this inability or reluctance to read and the resurgence of figurative art and sensibility-based abstraction? This characteristic manifestation of the turn of the new decade seems to me to function in a linked way, no matter what the specific histories of representational painting and sensibility-based abstraction have been in recent years.

*5 May 1979*

. . . The crude methodology of the student "crit" runs as follows: a rough sense of the orientation of the work is quickly grasped through a rapid examination — in some work a cursory glance suffices. Next, this rough impression is refined through a mental comparison against the production of artists who established the genre in the first place. The tacit assumption in all of this is that young artists are trying to be like a model of some kind; consequently, they feel pressured because they are not that model.

Many other wrong-headed assumptions — other than emotional harassment — proceed from this pedagogy, the most vexing being the assumption that young artists *confirm* the forms of a previous achievement. In part this is, of course, true — so much so as to be truistic. But the opposite corollary is also true — no matter the derivation or imitation there is also on some profound level a simultaneous *rejection* of the model roiling through the process. For to become the model presumes the young artist is there only to ratify a preexisting mode. Practical experience shows us otherwise. Rare is

the young person with such a mind-set.
. . . All art can do is signify. Signals are admitted of content, are said to have meaning, only through a certain sociological consent — though this society and consent occupy but small groups of individuals working in concert. The principle is also true of art put forth in the name of The People, though I am hardly talking about big societies. Indeed, "societal consent" on so vast a scale, on a national or racial scale, on the scale of *das Volk,* amounts to no more than a consensus-tallying of pluralities and majorities, attitudes that have scarcely anything to do with art but with concerns that are manufactured and manipulated, views that, in the end, are the enemy of art.
There is no meaning *in* an art object other than its own quiddity, its own material presence, that is, as mere neurological stimuli. This is true for a Rembrandt and for a Duchamp. All content-meaning is interpretive, imposed from without the object, from the outside.
Style represents the codification of a certain set of signals. I suppose I'm saying that style encodes and decodes these signals for comparatively small groups. Style also presents an ironical stance at the same time — at least for the group I'm referring to, since all art and all style proceeds from paradox. Style encodes — as does form — but neither style nor form is the encoding. Unless it is. Words.
When signals fail to obtain, when they grow faint, style is failing, is shifting and a new set of signs is being codified and beginning to be grasped. A sure index that style is changing occurs when boredom and depression accompany the art-making process, let alone the more public one of art-perception. This grimness, this negative experience, marks the sloughing off of one set of sign systems, one style, for the formulation and emergence of a new set, a new style; new skins for old, new lamps for old, new styles for old. Young artists feel this especially.

*5 July 1979*

. . . To Gerald Hayes' studio on Chambers Street — a bit of a nostalgic visit as they were Dorothea Rockburne's old digs.
Not especially interested in Gerald Hayes' newest work — apart

from its theoretical basis which is fascinating. The work is presented as complex *tondi,* addressing art as a language of sign systems covering Realism, Abstraction, Painting, Sculpture, Relief, Color, Minimalism, Expressionism, Photography, and so on — in short, real concerns, but isolated out — that comprise the critical issues of the '70s.

We both agree that the art of the late '70s is marked by these Complexities and Contradictions — thank you Venturi — but I am far from sure whether the current complexity of Hayes' new work makes it better than what had gone before.

Hayes is interested in complexity as it challenges and engages *harder making* as compared to the *harder thinking* of his earlier work — as if such a distinction were even possible to imagine or enforce. It's just that the earlier work *seemed* more conceptually felicitous, if not downright breezy.

Anyway, Hayes at last became turned off by all the facile art-making that he sees as having emerged from the Minimalist legacy. So he wants the art-making process to be more complex, harder, not because the art work would be better because of it (since judgments as to good, better, and best are not functions of sincere dedication or of difficulties overcome), but simply because the art itself would appear harder, more difficult of fabrication. Which it would be. And is.

Still, I am more taken by Hayes' work of the mid-decade. These are large photographic prints of plants, potted avocado plants, say, winning signifiers of modest New York apartment life (Plate 5). Hayes drew the circles implied by the double arcs of the avocado leaf contour — the leaf shape itself is to all intents and purposes the shape formed by the overlap of two circles — directly on the photographic print. Even more interesting than these drawings on photographs was the application of this work to wall installations. The photographs are pasted to the wall. The circles that extend beyond the perimeter of the photograph are drawn directly on the wall — engaging wall and print at the same time and becoming in this way a kind of environmental, perhaps even a para-sculptural installation.

There were other photographic wall drawings too — those based on spiky leafed plants — compelling the extension of long drawn lines extending from the photograph far across the wall; or hanging ivies (bunched circles); or all the "circles" implicit to the nude body. This last group of beautiful works looks a bit like Nijinsky's drawings of dancing figures made during his long institutionalization as a mental patient. Hayes' figure drawings also indicate an awareness of Oskar Schlemmer's designs for theatre costumes — how could Hayes not be aware of these celebrated Bauhaus theoretical designs? — not to mention the many Renaissance prototypes, the "Vitruvian Man" constructs, of which Leonardo's is the most famous illustration.

Of late, Hayes has been captivated by the image of two breast-like hills in Pennsylvania. The mounds are used to generate systems of circles from which in turn Hayes can postulate an entire index of abstract signs.

I like that one can treat nature in so blasé a manner. I misprize the sentimentalization of nature made by way of apology for art. Nature is natural. Art is artistic. Art is not natural. Nor is nature artistic. Never. I like that one can treat nature, naturalism, the landscape genre as just more grist for an abstract system. A Pennsylvania farm along a more or less vertical axis — like a connect-the-dot, follow-the-spot "house," say, in an 1880s landscape by Cézanne. Hayes' ascendancy — Cézanne to Johns (art as sign-systems) to Hayes. Not bad. But is it better? No. Thinking is not art. But if thinking is not art, then for sure feeling is not either.

*14 July 1979*

The intellectual core of contemporary French culture is essentially philosophical. Barthes, Foucault, Merleau-Ponty dominate, as well as a certain kind of Surrealist-Marxist-Freudianism such as one finds in Lacan. True, painting continues — "support/surface" —and it goes far in maintaining the untiring throttlehold of *tachisme* and the *informel* of the Fifties. A mixed grill of modernist conservatives — Dominique Fourcade, Marcelin Pleynet, et al., the French Greenberg distillates — still obtain in the critical sphere.

Truly vanguard intellectual art is to be found at the margin and a small dealer might survive there, but only with the greatest difficulty. At this extreme edge a critic or two of fresh optic might emerge but there is really nothing for him or her to review, to publically adopt as a cause. The conservative modernist writers have *Macula;* the *Tel Quel* group has the *editions du seuil*; they all have the ear of Pontus Hulten who might of a moment commission an occasional catalogue essay from them, but as director of the Beaubourg and after the suite of three magnificent exhibitions (Paris/New York, Paris/Berlin, Paris/Moscow), he has become far too cagey for the conservative modernists, throwing them an occasional bone as Talleyrand might have.

. . . The most disturbing thing in French art is its abject unquestioning acceptance that art is in some large and final way visual. Maybe it is but I hate the smug acceptance of the proposition for axiomatic. The Americans, to their eternal credit, have at least seriously doubted it and for long stretches at a time. Equally stressful in French art is the idea that scale is merely a function of the petty dictates of *objets de luxe,* transportable possessions scaled down to the size of seventeenth- and eighteenth-century rooms. Furniture and the salon — not the Matisses and Picassos that we have learned from — establish the perimeters of French art.

The truth, of course, is that this kind of attitude is to be seen again in American art of the later 1970s, in the resurgence of pretextual painting, "New Image," "Pattern Painting," "Neo-Suprematist art" — though the last named at least maintains a certain argumentative interest. And what of all the sensibility revivals of abstraction and figuration? All this attests to the deadening drag of an art presumed to be visual and whose premises are geared to visual delectation.

In the end, these are all part of a trendy and tedious rehash that marks so-called "post-modernism," an etiquette that allows for a great deal of delusionary posturing favoring plain old-fashioned eclecticism. That's okay, too. You can only do what you want to do. I might have it otherwise, but I can't make it otherwise.

Then there's me and that's a problem, my hedging, possessing as I do, both an extreme scorn for the traditions of the luxury object as well as a love for them honed by an appreciation of fine or prime examples of previous art, an art that, after all, never aspired to a conceptual basis in the first place — or at least we think they

did not.
The conservative modernists are always telling us in one way or another that it doesn't matter since epistemically-generated art in the end will be judged by standards or exemplified by "choice objects" anyway. Well, that means the goddamn visuality of art once more. I don't believe it, protest too much, refuse to accept it, am always frustrated by the ground-zero, the belief in visuality.
Might it not be that interesting art, at least in its inception, was the art that made the pretense, if not the real effort (and who can distinguish anyway), to overcome visuality? Or is interesting art only the art that embraced visuality from the outset? Could it not have been that interesting art was at least germinally aware that something of its meaning was anti-embodiment, anti-visual? Perhaps, even, this protestant germ may account for some of new art's initial effect. Even more, might it not be that a desire for, indeed a need, a yearning for anti-embodiment or anti-visuality turned in upon itself during the making process was transformed in making, so that it called down upon itself the strange appearance of ugliness, an ugliness that may be the strongest statement art can make against embodiment or visualization. For to embody or to visualize seem such paltry functions of the known, working solely in the sanctioned confines of received information. I don't know, but I'd hate my shrugging and walking away to be seen as merely the sentimental "out" that I suppose it is.

## ENTRIES:
## TWO VOICES

*2 April 1978*

There is a primordial content, a basic content to brute matter, to raw substance. But to say what this content "is" — other than itself — is impossible. More, that this "brute content" denotes an "art content" is disputable even when embodied in objects or things called art. The denomination "art," the very cognomen, is a source of its content — not its brute molecular existence.

Art content is not embodied in the object — at least in any way that can be elucidated — but in its having been assigned to a category. The designation imbues matter with certain value: but nothing can be said about "brute content" except perhaps to name it — if the name is known — and beyond that, perhaps to note that one may experience "brute content" in some way, as weight, say, or mass or color — experience it at some purely retinal or tactile level, as a function of our sheer neurological circuitry. But this is true of all matter whether or not it is said to be art. Thus, the "brute content" argument of a Rembrandt is exactly the same as the one for a Duchamp, or a shoe, for all that. The art is not in the object except through its assignment to a category — and the ultimate source of such assignments and designations still remains unclear although they are made an infinite number of times daily (and it would seem correctly, for to repeatedly assign things to incorrect categories is another way of describing the condition of insanity). It goes without saying that emotional response is a function of the experience of "brute content" of matter.

All other arguments, the myriad arguments (except that of "brute-content") that surround the art object are the art-content — the content of art is all of the arguments surrounding the sheer experience of brute matter. But sheer experience is itself not an argument; it is the bedrock. Art-content arguments are historically embedded functions of prevailing social values, the *Zeitgeist,* they are *dans l'air.* The attribution of argument, that is the making up of the content of art, is historically linked and sets up a chain reaction of seemingly unbearable equations (for the artist, at least): the content of art = the argument surrounding the brute matter of art = an historically embedded content = style, since style = the number of arguments imaginable at any given moment. And style, like water, like any natural resource or organic thing, gets used up.

*Style gets used up.* The artists who persist in "using up" a certain style are sanctioned in this exhaustion only by the fact that they were the ones who more or less synthesized the style in the first place. All late comers or diluters are not sanctioned in this waste. (Caveat: this is only an argument. Criticism functions in the realm of the general. Art — well, painting and sculpture — functions in the realm of the specific. Art = practice; criticism = theory.)

Is this why abstract artists of epistemic persuasion are angry with me (for the moment?) — because I hinted that their style was used up. Small matter — the inceptors of the style always will be seen as major figures. Their followers, modifiers of the style — if all that their synthesis amounts to is just another modification of someone else's given — will be viewed as little or lesser figures. Greatness — that heavy word — is only accorded to artists who create a new style — an achievement probably effected in concert with a still larger group of other, like-minded figures

*9 March 1979*

No amount of argument can ever *fully* explain anything be it a person or a work of art. The only argument that can fully explain anything is the object itself — *das Ding an sich.* Yet, since period determines argument, there are only so many *salient* arguments that one can append convincingly to works of art at any given moment — though how many one cannot say.

*6 April 1978*

Being in touch with a stylistic *dernier cri,* with the last word in style, is no guarantor of art (though it shows at least a sense of historical awareness and a certain gift for rapid cultural assimilation). Similarly, a dedicated commitment to earlier stylistic premises — essentially those of Renaissance space and Baroque tactile illusionism — is no guarantor either. Paradoxically, such intense revivals may be viewed, in their way, as ruses and political acts within the contemporary art world. The dedication to an earlier style — even an earlier "abstract style" — does reveal that there are broad reaches of artists who feel alienated from, cannot believe in, the art arguments of their own times. And in their alienation from such arguments they are motivated by them through what might be termed negative definition.

When artists revive earlier principles of style they show that there is something in the art of their own time in which they cannot believe. Such argument is, for all its backward-looking, a contemporary argument and possesses its own contemporary art-content — despite all the negative terms of reference. Among the things revivalists say (that is, through their art) is that they cannot believe in an art-content which they haven't invented (or so they imagine). This need to invent style forces a recapitulation of the history of art. They are saying that if one has not invented "it," one cannot believe in "it". And the revivalists possess no belief-system for ratifying the arguments of contemporary art-content. Mostly, this belief-system is a function of a close knowledge of the history of modern art. And for all their protestations, I find that so-called modern artists are often woefully lacking in a sure-footed sense of the history of modern art. So, I suppose, they didn't need to know it after all.

*11 December 1977*

There is a powerful connection between sculpture being the leading art of the Sixties and photography that of the Seventies: they both address the empirical fact. Sculpture is real, tangibly real; photography records, peels the skin off the real, steals its outermost membrane — the film of the real. Obviously they both trade in the literal effect of light upon object, of things illuminated or in darkness. One proceeds from the other and painting — that orphan of abstraction — is left out in the cold.

In our arts the avant-garde quotient is always a function of, indeed is measured by, the familiar, the known. Presumably, the audience for avant-garde painting and sculpture is acquainted with the conventions of representational art. Likewise, we assume that the audience of vanguard theater - the Dadaist or Futurist theaters, say, or the Pop Happening — is aware of the conventions and traditions accruing to a naturalist proscenium-arch theater.

*1 January 1978*

Realist art (and literature, too, for that matter) is marked by a sequence of highly specific detail, believable empirical detail, real data — yet each of the details of realist art must engage the spectator's belief in its ability to function on a symbolic level — each real element must suggest or infer the possibility of symbolic function — even if this is not the expressed intention of realist art. If the details of realist art are unable to invoke symbolic value, then they are not more than lists of things — sequences of mere literal happenstance — statistics into which one is hard pressed to project one's interest. It is the unspoken, frequently unwanted, muddying symbolic function of realist detail that transforms realism into art. This transformation can only occur as the result of an act of faith, primed by acculturation. It cannot happen unless the belief is there that it may happen. Hence, all "symbolisms" are mystical at core, impervious of analysis, functions of attributed content. This does not mean that the artist should strive for a symbolist look — nothing kills or cheapens realist work faster than an obvious striving the symbolist aura.

*16 April 1977*

Mistrust my own polemic against representationalism: if one can build with the theoretical template of epistemology, then one ought to be able to build with an actual model in nature. What I suppose I object to then is the ignorance of representational artists who work with nature because they know so little of the history of modernism, at least that part of it which could have provided them with the sense of an alternative ratification to nature. Representationalists behave as if there were no available ratification system to deal with the advent and development of the conceptual movement. Thus touch-based painting and sculpture is always there to function as the cultural bedrock. And because it always *is* there, it eventually comes to seem as if it were the only thing that ever *was* there.

*20 January 1979*

Crafts ratify the culture, are continuous with it. Art, by contrast, is incompatible with, disruptive of, the culture. Art is capable of moving the culture toward new paths. When (and if) this rerouting succeeds, then art begins to ratify the culture and in so doing becomes as crafts are, becomes crafts. The major artist is a demolitions expert always functioning beyond the culture's capability to absorb and to tame, to "co-opt" (as we used to say), to transform arts into crafts — crafts are socialized arts.

*26 January 1979*

A peculiar problem: the conceptual artist is in a bind on completing a body of work — since the subject of such an artist is inorganic to the hand (though it may be organic to thought and thinking), what to do next becomes the problem. With conceptually-based arts there cannot be the clear evolution built into hand-based, touch-based, touch-sensitve arts. With these the evolution of style is built, as it were, into the very muscle fiber of the artist. But what of the conceptual artist? Here there is no "next work," no "next step" built into the hands' development. Rather, the conceptual artist faces the search for a proper subject on completion of work rather than the mere maintenance of a certain range of muscular impulse and consequent retinal stimuli.

The effects of this dilemma have been enormous in the history of conceptual art. "Proper subject" or "good subject," as a photographer might say, accounts for much of the set back the conceptual movement has incurred recently. In periods of "conceptual retrenchment," touch-based arts look for all the world like the "winners," as if they had been there all along. And for a simple reason: touch-based arts never came to grips with the dilemma of the subject, always refused to deal with the advent and development of the conceptual movement. Thus touch-based arts are always there, a cultural bedrock. And because it's always there, it eventually comes to seem as if it were the only thing that ever was there. All this is bound up in, and inextricable from, the problem of representationalism. It too seems so solid, the most granitic statum on which we build our culture — for better or worse.

*1 February 1979*

Either/or: what is clear now is that my development as a critic in moving from a formalist premise to a diarist bias allowed me the luxury of possessing two voices. One is the echoing, mimicked voice of the artist whereby the critic soothes and ratifies — the critic as advocate. The other is the voice of the critic-as-adversary. In my writing there are now two voices present — the artist's and the critic's. In earlier formalist writing there was only one voice —one person really — the artist — though to say this of course invites the very natural protests of artists. They would surely say that the formalist voice was not their voice — but insofar as the formalist voice rejected privileged biographical detail and exacted the suppression of the critical personality, it was indeed the artists's voice, his surrogate.

## ENTRIES:
## SHEER GRUNGE

*27 January 1981*

To Martin Silverman's tight loft space on Greene Street with his wife Rona (first name here, despite heightened consciousness, as there is a work named for her) and their little baby girl Kate. The baby is kept out of Silverman's work space by a wooden fence. The whole place pervaded by a baleful Socialist Bohemianism, circa 1930. Silverman's at a tricky moment. In his career, I mean. Now "career decisions" impinge on "art decisions." Until now it has been grunge all the way. He's on to something and it could break suddenly, but in breaking could dissipate almost instantly.

Silverman's been picked up by the foresightful dealers. No absolute guarantee. His first one-man show for Ed Thorp didn't sell at all. Silverman's hope under pressure and his disappointment lead to *Crying Man.*

"I wanted to do something that was both depressed and positive," Silverman tells me. "So, *Crying Man.* It's OK for a man to cry. Okay?"

Silverman is a Williamsburg boychick formed by the proletarian left. He once even worked at Camp Bronx House in Copake.

"Okay, I feel better now," he says on learning that I was familiar with the settlement house camp though at a thirty-year remove. He played football and went out to Dakota (did I hear right?) on a football scholarship And from there to Hofstra, where, with Joel Shapiro as his teacher, he dedicates himself to art.

At Hofstra he makes his first sculpture, a figurative piece. The instructors are down on him for it. "Look, man," they say, "you better start doin' the art magazines." And he does. He becomes a '70s abstractionist - of linear, open, scattered, process-oriented sculpture.

"It was easy. It was fun. Bob Littman [then Director of the Hofstra University Art gallery] was really angry at me once. He was having an opening. You know how social those openings are. And I scattered a thick layer of clay dust over five tapes. I walked down the first tape, okay? I took a gulp of whiskey. Then I walked down the second tape, doubling my whiskey swallows, okay? I was really sick by the last tape. And the footsteps were all recorded. Littman couldn't clean up the floor in time for the opening. Boy, was he pissed."

Saw *Rona* again in Washington, D. C. at Diane Brown's new space in a show of Kirili, Saunders, and Silverman. No one will believe it, but my Entries of December 13, 1980 (*Arts Magazine,* February, 1981) that wrote of these three sculptors together was purely a chance connection. *Rona* is an elegant patinated spread-armed bronze figure. Of course, the word "elegant" is all wrong. Silverman aims at pure *faux-naif* klutziness — a kind of Thirties WPA feel, like the cornices in Parkchester combined with late Guston jokebook imagery gone clay therapy. But, still very personal (Plate 6).

In plasticene, the works are appallingly ugly. Silverman cannot afford to cast them yet. Hence, some of his frustration at not selling his first show. He couldn't go ahead with more casting.

There are a few ambitious pieces in bronze — *The Family,* which I saw at Diane Brown's, and *Run Around Sue,* which was shown at Pat Hamilton's Bronze Show. There's another plasticene dance group at the studio tentatively called *Passion.*

Silverman's first bronze is *Hunter.* It features a little duck, like a Benin weight, waggling at the end of a bronze wire in front of the striding hunter. The figure wears Gasoline Alley trousers and reminds one of R. Crumb's "Keep On Truckin'."

The patination adds immeasurably. The powerfully colored patinas are achieved through dipping the bronze in acid and chemical baths. They are then directly fired through open torchwork on the surface. "The easiest most lurid color possible is purple." To strengthen or brighten a patina a whitish layer must first be fixed and the subsequent chemicals fired over this. Without these white "grounds" patina color on bronze is dark.

Other works include *Martha, Diver,* a handheld mirror reflecting a clock (very *Crying Man* in effect), a man reading a book, a large girl legs akimbo with a doll, the dancing couple (*Passion*), and *Angels Fighting Hoods.* The Angels themselves wear two berets each, are Janus-faced, and three armed.

*5 April 1981*

SoHo in a state quite as if tom-toms were beating, the tattoo intensifying as one approached 420 West Broadway for the Julian Schnabel show installed jointly at Mary Boone and Leo Castelli. In the latter gallery, *The Exile* (Plate 7), a gold-ground antler-painting with the quotation of Caravaggio's *Boy with Basket of Flowers,* seems the most brilliant piece — despite a derivation from Rauschenberg's combine-paintings tantamount to homage. In Schnabel's work, the antlers seems as much the symbols of male genitals as do the stuffed birds of Rauschenberg's *Odalisk* or *Canyon,* not to mention the long-haired ram of *Monogram* — one is tempted to say "Monog-*Ram.*" Then, too, there is the boy-love Ganymede subtext present in *Canyon* (the black eagle of Jupiter carrying off Ganymede) reiterated as the homoerotic paean of the Caravaggio still life — *basket* of flowers, get it (Plate 8)? For me, *The Exile* is perhaps Schnabel's single finest work though the public, I hear, clamors for more "plate paintings" as that body of his painted assemblages is coming to be called.

The chief work at Mary Boone's gallery was *Painting Without Mercy* (Plate 9), a hugh Crucifixion-like "plate painting" à la Beckmann that interpolates quasi-Grünewald School gallows (crucifixes) with portraiture — William Burroughs among the notables — the whole thing very turgid and overburdened with the iconographic *horror vacui* typical of Schnabel at his most Burroughsish *Naked Lunch* aleatory. The label for the work crowed a bit, perhaps forgivably: "Collection Peter Ludwig." The "Ludwig Crucifixion" transformed Boone's small space into a chapel-like arena — Think Deep Thoughts — the sacred effect modified by the two large drawings hanging to the left of the huge altar-piece-like work.

As good as Schnabel's show was in itself, as well as for the artist, so too was it good for David Salle, whose works had been shown in

Boone's space just prior to Schnabel's. By now the conventional classic/romantic duality (emblematic for us in the work of Johns/ Rauschenberg) characterizes the Salle/Schnabel exchange. Salle, it is clear, is now becoming a deft painter whose progress points up incipient refinements and elegant turns of phrase, seen often enough in the ribbonish lines of the overdrawing, not to mention in sprayed soft-edge imagery and yielding fields of stomped figuration. The whole machinery moves ceremoniously, decorously, each element in place, no matter how complex the disguise sought. Salle now approaches the stately and the seamless, without disruption or disjunction.

Schnabel, by vivid contrast, is all gamble and impetuosity — rejecting at considerable risk all notions of accessibliity — flinging shrapnel about in his guerilla-warfare *cum* painting. All this "bad taste" suggests a career of giant unevenness — to moments when his pictures will work in the undisciplined way they do now, and to a time when they will fail on precisely these same grounds.

If nothing else, both artists are obsessed with style, past style, modernist style, no matter. One could almost say that a heightened consciousness of style is the raw material out of which they build their work. Only, to say such a thing suggests the ironic Pop tradition of art about art — and neither artist is especially ironic — though, if one were to point to irony, it is present far more pointedly in Salle. Well, at least, skepticism is. Schnabel, by contrast, is all about an unquestioned, regressive love affair with art.

What I like most about Schnabel's work is not the work so much but his sheer o'er-arching, knock-your-eyes-out folly. Would that this statement spoke only to the artist's sense of exaggerated defiance rather than to art, or style, or career. Even Schnabel, I suspect, must find his megalo-romantic theatricalized *kunstwollen* a trifle ridiculous.

But if he doesn't, no matter. Each new work of Schnabel's must be a trifle more "ugly" than that last — "ugliness" as sheer willpower — "ugliness," being for us, the surest way to avoid petty mannerism and *belle peinture* — that is, until we grasp anew that "ugliness" is (or can be) its own form of mannerism and *belle peinture.*

I recognize to what degree I personalize when I note my belief that all the detritus in Schnabel's work, all that broken crockery, thrown-out light sconces — all that raw materiel — are indices of rejected lower-middle-class life, its sheer grunge, as it gets lived out in vast stretches of Brooklyn, Queens, the U.S.A. really. Then, too, all the smashed crockery is a scaled-up unifying field that the Cubists invoked when they pasted down newsprint — the cracked saucer in Schnabel's case prefigured in the cropped headlines and graphic subtexts of the Cubist collage.

The bridge, then, between Schnabel's china and Picasso's *journaux* is perhaps not only Rauschenberg's combine-paintings, but the disguised newsprint submerged in the waxen fields of Johns' encaustic paintings of the mid-1950s.

But we have forgotten how theatrical Picasso's choice or Johns' choice once seemed — so classical have these oeuvres become for us. By contrast, Schnabel's protracted-adolescent self-dramatization seems vivifying, "wow," risk-filled in a way that allows for a shared exuberance we are perhaps only too happy to lend ourselves to, a certain self-deluding collusive participation.

# JULIAN SCHNABEL:
# BLIND FAITH

## I. The Studio

*1 September 1981*

Talking to Julian Schnabel is a bit like playing snooker: the cue ball scatters the others — some line up sensibly, the rest just go any which way. Schnabel's new life isn't quite so agreeable though it sure sounds it. Reckless comparisons multiply — to Pollock, Rauschenberg, Beckmann, Picabia (Plate 10), whomever, that sort of thing. The snowballing has been hectic, as every reader of art magazines knows.

A few months ago, Schnabel moved into a more or less finished working space — sliding panels, bright responsible young assistants, phones ringing, stacks of carpentered stretcher supports, piles of Le Franc oil paint in pound tubes — an incredible dream of luxury, a monument to painterly orality; in brief, the extravagant appurtenances of overwhelming success. Despite the frenzy, there remains the reticence of an artist who wants to be taken seriously even in the face of a worldly success and reception that is frightfully off-putting, even for him.

It is easy to suppose that Schnabel's art resides in the associative eccentricity of the substances chosen; for sure, there is that. Earlier, one thought that Schnabel's shtick was just a function of an elaborate personalistic iconograpy; for sure, there is still much of that. Earlier, one "justified" Schnabel's pictures in terms of a formal continuum with the Combine Paintings of Rauschenberg — appealing to a legitimization afforded by the history of collage and assemblage. Then, too, the "transparencies" of Picabia (from the 1920s) as well as Picabia's "vulgar" calendar art images of the 1940s also provided models. True, there are all of these resurgent iconographic parameters, but now it seems that Schnabel's major impulses are being realized through sheerly assertive surfaces, surfaces that either invoke ironic signifiers of low-class values (the velvet ground as a reminder of "genuine hand-painted velvet ties"), or specific cultural invocations of powerful ethnic identities even if, albeit, "colonized" ones.

Huge rolls of outdated kitchen linoleulm — signifying marginal apartment life, say — await their transformation into supports for paint. New pictures are painted on grounds of broken Mexican ollas and clayware or short-order souvlaki amphorae, desert buffalo horns and leather, and the like. Still, with regard to these new works, by no stretch of the imagination can Schnabel be taken as a Mexican or Greek or laboring-class apologist. Indeed, the agressive presence of these cultural artifacts, if they can be said to "function" at all, do so from a position of ironic indifference — Degas and the laundresses — subject matter viewed *de haut en bas.* Schnabel is hardly a popular propagandist, though acutely he has no aversion to his work being perceived along these lines. And in the end he cannot control this anyway. Anything ever said or written about art is its content — though maybe not its "art-content."

So, do not be misled when Schnabel says that "art is not painted for an informed public . . ." He certainly does not mean by this the popular ethnic illusionism of the *sans-culottes.* His corollary quite properly continues that "art is received by an informed public."

Schnabel's primary formal concerns address the question of how to apply paint to disruptive and "sociographic" surface — art for art's sake. All of his "iconography" or "iconology" is pretextual. Nor is this necessarily a failing. Such matters are purely neutral.

Take, for example, those clay elements dispersed in plaster. Schnabel paints for all the world as if there were no disruptions at all, as if the support were smooth. The brush turns only when it bumps into some large blockage so that it can no longer assert its original directional impulse. There are constant dual, even triple, readings: a pretended uniformity of surface countered by the ructions occasioned by the relief elements. There is the difficulty and disagreeableness of illegibility of image, the disjunctive and near-haphazard character of the work; then again, there is the conjunctive character of a visual intelligence which is capable of mixing — as an act of will — two layers of visuality. The "support" may more properly be seen as a kind of "depth of field." And painting becomes not an action on a surface, but actions perpetrated within an arbitrary relief-like corridor of space.

Though the fields of broken crockery are in some measure composed during their initial placement in plaster, the final compositions do not necessarily derive from these placements. Often, the painted parts of the compositions run counter to the natural placements of the relief elements, indeed violate them.

On occasion, an exchange takes place between Schnabel's observation of an ornamental motif (found ready-made, pre-existing upon some straight element). He then might visually isolate it and stress the subsequent rendition as a prominent pictorial element of the newly painted composition. This expanded, blown-up decorative motif, image derived from isolated pattern, is particularly characteristic of the new "Mexican" works, *The End of the Alphabet, The Beginning of Infinity* (Plate 11), say.

This attitude of extreme disjunction — big to small — is a working method also seen earlier. In the "drop cloth paintings," enormous painters' drop cloths - what else? — were placed on the ground Upon these large surfaces very small images — say, the famous self-portrait of Antonin Artaud [one of Schnabel's heroes (reproduced from a photo in the *Paris/Paris* catalogue)] — was scaled-up by a rendering of minute features with three-foot brushes. Or, "big to small" might be occasioned by Schnabel's examination (while it was held in one hand) of a small primitive or religious carving, a Santos, say; with the other, he rendered this small devotional object with the objurate and reluctant tool of a three-foot brush. But it was the process that counted, the way of deriving the image, not only the resulting image.

Granting conditions such as these, it would be absurd to anticipate a cohesive, seamless spatial illusion and one doesn't. The space depicted becomes in a manner of speaking — the artist's speaking, at least — "indigenous" to the material (as well as one invested in the image); or "indigenous" at the very least to the wide discrepancies between depiction and that which is depicted. This need not be surprising. Artists' talk has long sanctioned, say, the notion of color as indigenous to shape. Why, then, not these other "indigenous-es"? Schnabel speaks with remarkable acuteness, though his sentences are so finely honed and deceptive that to miss a word is to often gainsay meaning. Of the processes noted here, he observes that as his work "becomes a painting, it becomes interesting as painting."

## II. The Loft

*10 October 1981*

Schnabel doesn't paint where he lives anymore. The works at home are now the pictures of changing date that may have come back from an exhibition, or may be en route to yet another, or pictures he has decided to keep, or new work that he is just thinking about. Indeed, the work is here mostly to think about — where it came from and what did it point to in terms of the artist's very rapid evolution.

An extravagant spacious disorder about him, a growing collection of unmatched furniture by pioneer designers of the 1920s through the '50s — Breuer, Aalto. A more recent acquisition — Max Beckmann's portrait of Wolfgang Frommel.

On the wall hangs the old picture (well, 1979), *Procession for Jean Vigo* (Plate 12), just back from somewhere. The T-form, that gallows/crucifix shape so often seen in Schnabel's work, derives in this case from a memory frame of Jean Vigo's film *Zero de Conduite*; while the motif resonates with Danube School crucifixion memories (as do very many others), the structure more specifically images Schnabel's memory of marauding school kids' "gallows for dry fart," that strange frame with lanterns dangling from the crossbar that the school kids erected to torment the sleeping monitor. Of course, its triune associations are inescapable and its Germanic symbolism invokes as much of Schnabel's fascination with primitive crucifix imagery as the antlers used in many of his combine paintings invoke the legend of Saint Eustace in which an apparition of the crucified Christ is borne upon the antlers of a stag. Then, too, the loping interlace, the triple helix, updates Picabia's *LHOOQ*, itself a paraphrase of Duchamp's celebrated altered ready-made.

In the *Procession for Jean Vigo* there are just some holes sprinkled about the picture. "The holes are there to disregard," the way all the relief detritus in the more recent painting is "just there to disregard," as if one could. Then, too, granting that you can't, it orchestrates its own system of cryptic association.

Schnabel speaks of this meandering conflation as the meeting of "different temporalities," a contemporary fascination with "simultaneity." But it is hardly the "fourth dimension" of the Cubo-Futurists, the introduction of a specious sign-system for time, that Schnabel aims at. Instead, he hopes to stimulate a sense, perhaps, of time lost, of introspection and meditation.

"Different temporalities," then, may be likened to those simultaneously presented frontal zones one finds, say, on the walls of Italian churches. These fragmented discontinuous layers of image and relief incorporate unexpected juxtapositions and associations: a Baroque memorial tablet fits into fragments of late medieval fresco edged by plaster and exposed brick and upon which may be pinned votive offerings of all kinds; not to mention the vast religious paintings from the Renaissance through the present. This, too, presents a kind of simultaneity rendered reasonable owing to our acute senses of history and connoisseurship and the renewed vigor of our visual appetite.

Schnabel wants to speak so simply and reduce his creative experience to as clear a phrase as possible that often his sentences seem cryptic or evasive; in fact, they are straightforward descriptions of precisely what he is doing.

Remember the holding of the small object in one hand and the recording of its features with "inappropriately" scaled tools on "inappropriately" scaled surfaces or colors? This notion is continued when Schnabel says that "my shapes are found by concentrating on another shape," by a kind of "looking here but painting there. Blind Faith." In this way, "the shapes arrive at themselves." And they do so through a transformation in which individual status may be suppressed; or, as is also the case, rather confusingly imposed. "I like shapes, simple shapes. I like their breaking down and their breaking up. Like torsos. They are like poplars or mummies or trees . . . ." Now Schnabel is free-associating: "They relate to nature, to leaving off a line, to cutting a statue in half, where the veins in the leaf become cracks.

This elaborate synecdoche contrasts not only legitimate parts of the wholes — for example, leaf for tree, tree for nature — but it obtrudes as well onto yet more extensive areas of substitution, the leaf for statue, say.

Still, Schnabel's intention does not favor ambiguity. If anything, Schnabel dislikes ambiguity as being a place "where artists hide out." Instead he seeks clarity and concentration of emotion, despite the immense ambiguity of the image one is given to digest. What the emotion is, of course, is difficult to pinpoint; but for Schnabel it represents a "certain kind of feeling, the feeling of a certain kind of claustrophobia."

I am reluctant to allow Schnabel to fob off what all too easily might pass for a sentimental piety. I press him to be more exact as to what he means by "claustrophobic feelings." Schnabel would have us see the work not only as its inescapable material presence. Despite all the perhaps excessive hyperbole of his work, he would have us "remove the painter from it, remove 'me' from it. All the works are made to be seen without sensing the artist around." Then maybe you can get what he senses as an ambivalent exchange between "intellect versus madness, a flirtation with the acceptance of death."

One sees how easily, from this viewpoint, Antonin Artaud could become the culture hero he has to Schnabel. The artist wants to retrieve for painting "the idea of death, the void, the abyss. We are responsible to these unspeakable things even if we cannot name them." But Schnabel acknowledges fully that "what I say may be very different from what one gets from the painting."

# DAVID SALLE: HOLIDAY GLASSWARE

*3 December 1981*

Complex David Salle — lean of face, tense, dark hair on a sharp cartilaginous profile — the unflinching gaze of the contact-lensed. A cigar-smoker (by way of affectation?) and a just-audible William Buckley-like speech pattern. Earlier on, I was unduly cautious (as we all were) of what was perceived as Salle's careerism (and Schnabel's, too, to be fair — all of "New Wave" careerism really); while fascinated, I modified support in ways that emphasized Salle's critical/analytical faculties — what he calls, self-deprecatingly, "my criticality, such as it is." Recently, when asked what the "ism" of the '80s would be, I replied "Careerism." While lecturing at the Hartford Art School where Salle once "stinted" as instructor of painting, I spoke to Salle's paramount critical/analytical gifts, scanting in so doing his authentic role — that of a painter born. So in Salle's case, the moral of the tale no longer applies.

The Salle studio is in that row of buildings in which Stanford White — shot by Harry Thaw — died; raking views of the Flat Iron Building, Madison Square Park, and the Metropolitan Insurance Tower. Originally a living loft for Salle and his wife Diana (a coincidence, that, the Diana of Madison Square), the duplex (anomalously redone in the '50s) now serves as painting studio, complete with a *de rigueur* young assistant. A spiral staircase, rooms furnished with examples of '50s furniture once more *en vogue*, a pair of pickled oak fake-Archipenko lampbases, works in progress, finished works being wrapped in plastic sheets about to be sent hither and thither: large rectangles, powerful erotic shots, isolated abstract images, stomped passage-work, cartoon-fragments, scumbled impulsive areas, varied-scaled abutted rectangles, some colored, others not.

First formulations of arguments that trailed away. Salle has extraordinary gifts for cautious postulates so considered in their turning that they are apt to simply fade: "Conceptual art allowed my painting to exist . . . ." "My work is simultaneously itself, its own representation, and the idea of itself . . . ." "I am not really interested in the referent, nor to deny the reference — but I am caught up in the way the works are left to stew without being watered down. I am interested in the specificity of their presentation."

Salle's not talking so much, I suppose, about what the image depicts, but somehow of the gravity, the weight accruing to the image. "And I don't see my work as some kind of bridge out of Pattern Painting to Punk, though there may be patterns; I see my painting as emphasizing a dysfunctioning network of references that establish possibilities outside themselves." Of course, Salle is right. So aware is he of his "critical intelligence," he could allow for the ambition to gratify a functioning system of inferences. After all, "critical intelligence" establishes the logic of references and the perimeter in which these references function.

Still, as it happens, Salle's painting also appears to embody the entire opposite of this ambition. "Sign, symbol, metaphor — all that only matters insofar as one holds it together all at once at the same time. Never losing sight of any of the permutations, the overtones —that is the nature of the experience I want." In short, Salle's imagery is never resolved, only "over" when one turns one's glance away.

Nor is there an aesthetic hierarchy of images; nor one of touch. One gesture serves as well as another; one color as well as another; one impasto frosting as well as another kind of surface; matte or glossy, small or large, sexy or bleak, cliché'd or novel; none of this matters except as it functions within a potential for constant conflict. And then, only as it matters in the most conservative way imaginable; that is, only Salle's work can be sexy or bleak in just that way: or granting the chameleon-like myriad options of touch, still only Salle would make the touch that way. So, in the end, despite the myth, the vehicle does remain *faktur,* touch, only his touch, and, in just that way, in just that place.

— "Why are you so combative? Why is your art so contentious?

What is being defended?"
— "My sensitivity. That is always under threat. More, the specificity of my sensitivity is always in danger."
— "Are we speaking of the defensiveness of the aesthete who, in order to hide, is often a most public chameleon?" It seems the psychological terms of Salle's strategies contrast the chameleon-like aptitude of the aesthete with what passes for the engagé, in present terms, little more than a code word for slob. All that's mental, of course. The outside remains urbane if a mite dour, a logo on a split-timber fence.
— "What is the counterpoise to the sheer selfishness, the egoism implicit in the aesthete's désengagé position?" I wonder aloud. (Probably exhibitionism and narcissism, the essentially untreatable inversions or denials of the necessary masochism built into the very neutralization of the aesthetic condition.)
Salle comes from an unexpected place — Wichita; Dad from Ohio, Mom from Oklahoma. Salle's escape was through art. That's banal enough. Rescue came in art school: William Dickerson and Karl Nordfeldt, characteristic painters of the broad Art Institute of Chicago manner of the 1920's. His folks wanted to be "modern," did the house up in what '50's suburbia was meant to look like, rented abstract paintings from the local museums. All this and more — the touch, the imagery, the anomaly of the suburban '50's — concerns the "new painting." In Salle's case, its pictorial terms are all rendered perfectly.
Fodder for the New Art — not just Salle's (and to be fair, rather little in his case) but an entire generation's come of age — Suburban Beautification. Looming behind are the turreted Hansel & Gretel castles of Disneyland, Disneyworld — orange & aqua plastic shakes, prefab distress and weathering, the Howard Johnson Shingle Style. Daylong, lifelong exposure to TV cathode rays have burned out the color cones of the eyes; naunced discretions are in the balance, must go, have gone. Smart conversation as sitcom one-liners, gags or the dutiful, portentous banalities of the soaps, six words strung in a row if even that. Art imitates Life. "Bad" is good, better, best.

*3 February 1982*

Of course, the Maximalist New Wavers Down at The Clubs are all inventing art out of the conventions of middle-class popular culture. Minimalism (epistemic abstraction and sensibility minimalism included) was perhaps the last cohesive style to emerge from the stimuli of High Art — the art of the text rather than the TV sitcom. And as is always the case with fresh cohesion, symbiosis is in the air. Though, after the fact, one sees that the emergent individuation was present from the beginning, especially in artists as challenged as Salle — or Schnabel.

"We projected different sides of one thing, not that we shared it." "It" remains a vague intangible of concern to the cheez whiz values of the soaps or the more sinister appeals of B-movie stills and the films noirs of the 1950s and '60s.

"Julian says to himself, 'I can do anything I want,' an essentially egoistical position. The result is a body of work that is radically uneven." Salle's perception is correct but one that obtains not only in Schnabel's case but in all o'erarching Romantic effort. These always call for heavy pruning, careful editing, to achieve a body of work that appears even-textured, convincingly seamless. The injunction obtains in Salle's case too.

"For me," Salle goes on, "the meaning of style is *use,* the instrumentalization of style." He differentiates this from "the instrumentalization of the motif," one which leads to "iconography," however neutrally or indifferently employed. Iconography, of course, indicates choice — no matter how unconsciously motivated that choice might be. The appropriation of image, the current despoliation of art and culture, always invokes the specter of specific interpretability. If only one can find the key. Once it was provided by psychoanalysis. Such procedures seem frighfully outmoded today. Marxist dialectics? Frankfort School phenomenology? Post-structuralist semiology? Yawns.

Salle promotes his case as "an instrumentalization of style." This leads to "the meaninglessness of touch," an irrelevance that impinges on "the parameters of painting or something called painting." Salle's distinction between the instrumentalization of style and that of motif is meant to point up the difference between

his work and Schnabel's. It all is purely theoretical, as there can be no mistaking the two anyway. But his view of the matter allows him the comforting belief that his work maintains modernism's dialectical shuttling, one that runs counter to a fixed Expressionism that demands no internal dialectic for combustion — Schnabel's art, say.
The whole mechanic seems too pat and programmatic. One has been there before. Substitute terms. Classicism (read ratiocination, dialectic, modernism) versus Romanticism (read revivalism, Expressionism, stuck). Useful, yes, but too wanton and too specially plead. They both are wonderful. They both know it, and none of this says anything about art. Individuation.

*17 December 1981*

Continue to talk to Salle at the studio and ask, "When do you know the work is finished?"
— "When what I have talked myself into for two weeks falls away." Then, again by way of locating him to his Cal Arts Baldessari-like origins, I questioned him about the role of the conceptual in his painting.
— "You don't have to paint. To paint is a piety I would like to keep alive, though the culture probably won't let me." It has, what with the unquestioned return to painting all about us. The canvases in the studio present startling discontinuous abutments of images spurred by a highly personalistic choice. There is a stomped, Jackie-like head that looks derived from the notoriouis image of a Vietnam execution — but still Ron Galella for all that. (Actually it is from a French photograph of children bound to a tree.) This image counters a broad, sparkling dark canvas based on Derain (Plate 13). Another canvas takes off on Félicien Rops — the coquettish domino-buttocked woman; a large older photograph of a nude woman with dunce cap on head and two cones extending out from her breasts awaits being incorporated into a still larger composition (Plate 14).
We are sitting in front of a large still life still in progress, based on the slick sparkling newspaper illustrations published by Lord &

Taylor in the *Times*.

— "I was drawn by the malevolence of cut glass. This panel is painted on photo-sensitized canvas. You make a mark on it, but you can't make a recognizable image because of the photo-sensitivity. And if you make too many marks on it, it all goes black. Then, I started to paint the look that rhymed with the newspaper illustration — Holiday Glassware" (Plate 15).

— "To be honest, David, I don't really think you successfully caught the slick skills of the department store illustrator; so as a diversionary tactic, you added the little green figure, a pentimento, a throw-away, as a kind of deceit in the margin. Of course, it doesn't really matter that you are not a skilled newspaper illustrator, since, after all, it all ends up as 'just painting' anyway."

— "You know, I have this revulsion of saying things over, and I act out my dissatisfactions with the picture as I paint it."

— "I admit that the failings of the passage are obviated in the very act of describing the scenario of the painting."

— "But," Salle interjects, "my painting is not about Abstract Expressionism's 'natural error.' There are fundamental distinctions between verbal and pictorial recognition. Because I got to it emotionally while talking to you about it doesn't mean I knew it visually when I painted it. Everything that is felt or perceived can be verbalized, though," and here we both agree, "the verbal only may be — probably only is — a descriptive catalogue without affect."

1

2

1. Eric Fischl, Critics, 1976. *Oil on glassine, 73" x 128".*
2. David Salle/Julian Schnabel, Jump, 1980. *Acrylic on canvas, 84" x 120".*

3

3. Alain Kirili, Alliance, 1982. *Forged Iron, 67″ x 65″ x 72″.*

5

4

4. Eleanor Antin, Eleanora Antinova in L'Esclave, 1980. *Photograph.*
5. Gerald Hayes, Philodendron — Every Arc, 1974. *Temporary installation: ink on photostat and Wall, 9'6" x 13'2".*

6

6. Martin Silverman, The Businessmen, 1981. *Cast bronze, edition/6, 41" x 19" x 14".*

7

7. Julian Schnabel, Exile, 1980. *Oil, antlers, gold leaf on wood, 90" x 120".*

8

8. Robert Rauschenberg, Canyon, 1959. *Combine painting, 81¾" x 70" x 12".*

9

10

9. Julian Schnabel, Painting Without Mercy, 1981. *Plates, putty, oil, and wax epoxy on wood, 120" x 168" (four panels).*
10. Francis Picabia, Untitled, 1930. *Collage, 12" x 9½".*

11

12

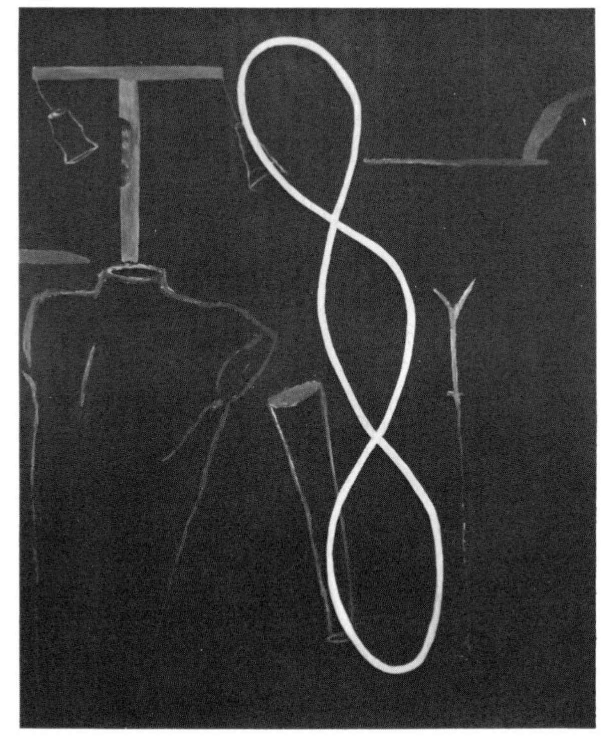

11. Julian Schnabel, The End of the Alphabet, the Beginning of Infinity, 1981. *Oil, pots, antlers and plaster on wood, 108″ x 156″.*
12. Julian Schnabel, Procession for Jean Vigo, 1979. *Oil on canvas, 110″ x 96″.*

13
14

13. David Salle, The Old, the New, and the Different, 1981. *Acrylic on canvas, 96" x 150".*
14. David Salle, Autopsy, 1981. *Acrylic on canvas and photosensitive linen, 48" x 112".*

15

15. David Salle, Holiday Glassware, 1981. *Acrylic on photosensitized linen, 91" x 91".*

16

16. Robert Longo, Corporate Wars: Wall of Influence (detail: center), 1982. *Cast aluminum, 7′ x 9′.*

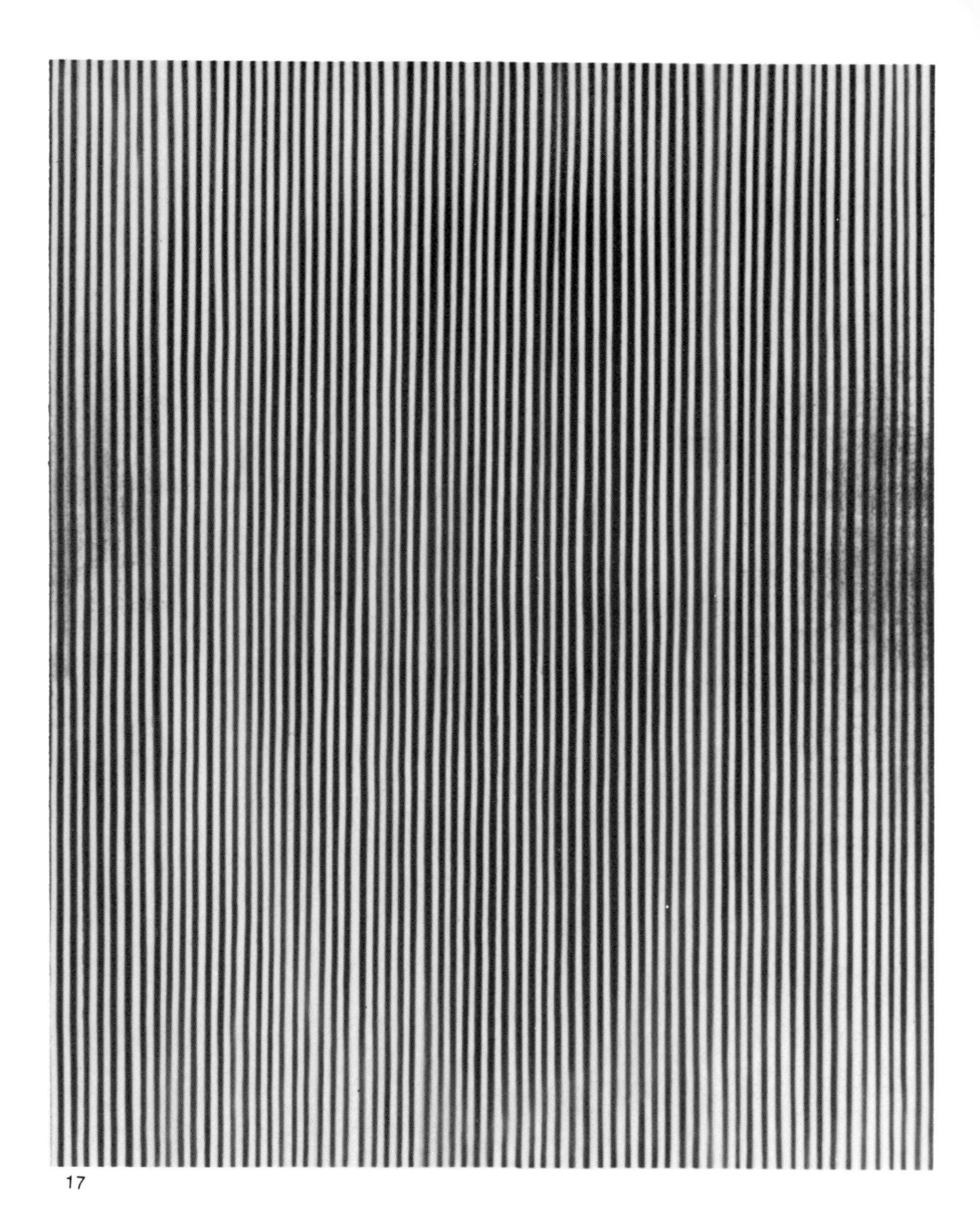

17. Ross Bleckner, Untitled, 1981. *Oil and wax on canvas, 120" x 96".*

18

18. Gary Stephan, The Agony in the Garden, 1981. *Acrylic on canvas, 90″ x 48″.*

19
20

19. Gary Stephan, Magic Motor Boat, 1965. *Oil on masonite with metal, 22½" x 61".*

20. Gary Stephan, Untitled, 1970. *Polyvinyl Chloride, 42" x 96".*

21
22

21. Gary Stephan, From the Athenor series, 1973. *Oil on canvas, 50″ x 106″.*

22. Eric Fischl, Bad Boy, 1981. *Oil on canvas, 66″ x 96″.*

23

23. Eric Fischl, Time for Bed, 1981. *Oil on canvas, 72″ x 96″.*

# DEFENESTRATIONS:
# ROBERT LONGO AND ROSS BLECKNER

*4 March 1982*

Imagine, people believe in the difference between Modernism and Post-modernism. All bright young sets claim a difference from preceding gospel — it's truistic. Modernism is an open caliper taking the measure of all within, even Post-modernism.

Still, that young artists may feel, do feel, different from past artists is important. If they believe in their peculiar focus, say the way the Metro Pictures group refers to their receptivity to an awareness of urban terror, well, fine, even if the dread they point to has a venerable name, paranoia.

The sense of differing consciousness, hence of altered choice, that young artists mouth as a matter of course, and ultimately promote, is a component of all style shifts. *Plus ça change . . .* to invoke the old saw is no waving a flag of obdurate hostility or indifference (as is often the case when an oldtimer appeals to the jaded view). Indeed the belief in the difference is thrilling even as the difference has all the distinguishing features of a conventional avant-garde posture.

It seems little different if Minimalists point to Suprematism as their paradigm and the Metro crowd points to B-movies, films noirs, or TV sit-coms from the '50s. Both groups remain the same in that they seek (or is it that it has been sought for them?) earlier sanctions - the one in high art, the other in popular media. In any case, an acute awareness of difference (even when unfounded) does effect style change — if only in this linear way.

*31 March 1982*

A visit to Robert Longo's loft on South Street, near the Seaport. The loft somehow hangs on even though all around, the city is being dug up or restored, archaeologically excavated. The loft was taken when Longo was part of the Hallwalls crowd (working out of Buffalo) — so many of whom have now begun strong careers in New York: Cindy Sherman, Michael Zwack, et al.

Longo is more rumpled than I imagined he would be, a genial dumpiness that the '50s d.a. haircut does not belie. The Kosuth-like black shirt and pants don't ameliorate this effect. The loft is tightly organized, revealing a tough-minded professional sense of what his work is about and how it will be promoted. Fliers for a yet-to-be-completed movie are available; photographs and slides are quickly accessible. The tall 8-foot-high drawings upon which his first success was so strongly built are set about, taxing in their close observation as in their execution and very impressive.

Longo's new ambitious sculpture is made of plasticene awaiting its casting in aluminum, perhaps bronze. The plasticene substance is less repellent than usual as the arms and faces of the enormous figure composition have been polished, shining, with a water wiping — indicating a contrast between glossy and matte surface.

The relief (Plate 16) is astonishing. About 12 feet wide and 8 high, it fits snugly just below the low loft ceiling. The figures crowd (some 15 or 16 of them) and are engaged in a frozen-time fall and struggle as if they had been "lifted" or extracted from a photograph or film frame. The relief, complex and technically demanding with passages often in full relief, invokes civic architecture of a conservative stamp, of a kind once conventional to academic sculpture. Indeed, ironies about academicism are endemic to this *Real Life* magazine mode. References range from motion pictures to the early Michaelangelo battle relief, the Raphael and Caravaggio *Deposition* figurepieces. Not just a lucky guess — Longo mentioned them in conversation.

He has been sweating, at work three months now without break, in order to finish the model in time (the very next morning), for it to be sliced into thirds and cast for a photo-finish inclusion in this year's Dokumenta 7.

The relief, wildly virtuoso, is based on a neighborhood stroll — street smarts — Longo often takes to the Stock Exchange. There he watches the daily Battle of the Business Men as they violently elbow

one another on the Exchange floor. It became Corporate Man's Battle of the Centaurs and Lapiths - a fascinating spectacle Longo views from the spectators' gallery. Baroque warfare. Many extravagant passages — a female figure borne aloft in the upper left; a gesticulating kneeling figure in the left; a business man duplicating the Christ of the *Deposition* in the center; the double vertical freestanding arms of combatants in the lower left; violently groping figures wrestling slightly to the left. For sure, spatial and anatomical impossibilities abound and in their excess a Mannerist model is alluded to.
A paperback of New York architectural sculptural ornament is open nearby for easy reference. Longo acknowledges his appreciation of Noguchi's Associated Press Building aluminum relief in Rockefeller Center.
Longo's battle relief is to be set into complex flanking elements — Deco-like views of raking office buildings. The violent perspectival foreshortening suggests not only the Art Deco, but also views of buildings as if seen by someone falling from them.

*3 June 1982*

Visiting the hyperbolic Ross Bleckner. A fascinating morning marked by the artist's brilliant rushes of insight and arching metaphor. Bleckner is a crisp assessor of the scene. His art begins to impress itself, to succeed, though often à *rebours.* His is a painting (and a painterliness) that, in a certain sense, may be said to want a certain naturalness, an ease, a fluency. Of course, I'm wrong. It's not that Bleckner wants a gift for painting (he has that in abundance), but, rather, one for color - a quality, not incidentally, characteristic of most painting today. So, in this absence of a natural ease with color, Bleckner is by no means unique. It is a reason why his work reminds one, say, of Edvard Munch who proves anew that one can be a great artist without, in fact, being a great painter. Do I have to flail you? the difference I'm drawing is between art and painting. Beckmann's awkward grandeur often shows this, too.
To note these comparisons is not to suggest that Bleckner is anywhere near genius — it is simply to make a point that obtains as well in the cases, say, of Pollock, Schnabel, or Salle. None of these

painters has a gift for color, but for painting — yes. By contrast, Stephan, say, has a gift for color.

Bleckner, 33, can be witty about the literal situation in which he finds himself: "Look, I had to do painting with bars because I lived right next to the jails; and I had to do New Wave painting because I live over the Mudd Club."

Since Bleckner lacks (presumptively) a genetic coding for color in the DNA spiral, he craftily transformed anemia into strength, addressing himself to painting in black alone and/or ironic commentary upon Op Art — yes, Op art — yet another 1950s signifier and middle-class revenant that young artists are exploring today (Plate 17). Of the blackness of his work Bleckner notes by way of countermand: "When I close my eyes I don't see color. Yet my black-and-white paintings are completey infused with color. They're tinted."

Bleckner's "opticality" also disguises a verge-imagery which just may be there, perhaps isn't. "Each painting randomly mutates. Op is a kind of failed imagery. The pictures appear to have iconographic interest, but then don't. There's a level of randomness. Simultaneously you want different things to happen."

This multiplicity and ambiguous richness struck him as a potential at his most recent show but one. Instead of a monolithic Minimal clarity, "I wanted the show to suggest at least five different branches of meaning." With the Black Paintings, Bleckner realized that "I could have what I wanted and I could have what was supposed to be identified." The earlier work, to be sure, had been marked by a stress on specific image. The image bank is Picabia/Man Ray — that is, New York and Paris Dada, especially Man Ray's *Revolving Doors.* "So now, at least, there's an image. The Black Paintings are a kind of veil, a layer leading to an afterimage. Only it really doesn't mean afterimage."

"How can I have just one thing? How can one image really suffice?" the artist queries. "The problem is to define yourself through your production." Hence, "A lot of artists spend a lifetime to find this equivocal image. But no sooner than found, it 'deconstructs' [destroys?] itself — as it really couldn't possibly be a carrier of so rich and important an intention. Iconic abstraction can't be glommed onto iconic experience. It's just too complex."

"Moreover, my sensibility is essentially not a formal one. My narrative is personal and disjunctive. The objectives of one's scrutiny don't make themselves known in a logical sequence" — this perhaps by way of refuting a widespread belief that all painting can be seen as possessing logical verbal meaning.

In these ruminations and self-doubts, these expressions of a belief in an important newness and existential enrichment, Bleckner shows himself still to be a committed and elemental Modernist. Against all reason, his is perplexed by Quality — that bugaboo — for without "quality," credible transcending presence, Bleckner's tantalizing self-doubt would be quite beside the point.

Conversation, though easy, ignited when I spoke of artists whom I saw as emblematic of specific movements because they are so stylistically hidebound in a way that Bleckner's diverse intentions would never allow. It matters little whether they continue to grow (as indeed they give every indication of doing) as their essential contribution is measurable in their determining relationship to the sense of the immediate. The point of their work is to be frozen and stabilized. As artists, Bleckner called them "Open Signifiers." (This is in no sense a statement of quality though I can see the horror registered by readers of these notes. I am the culpable party, not Bleckner.)

Bleckner and I were strolling now with mutt Renny (Renaissance), partaking of the noontime sun with the office workers of his Foley Square municipal neighbors. Bleckner carried "open signifiers" further. Referring to those artists I had casually savaged, he continued that they (and many others) were artists "whose work was immediately subsumed into the conventionality of the discourse."

Bleckner is, of course, ambivalent about his own rejection of, or anxiety over, meaning — assuming all the while that the "quality" of the paintings nonetheless warrants this search. In his evolution Bleckner identifies three central events, *just* events. The first is *just* becoming an artist, *just* going to college," to NYU Washington Square where, it so happens, he took a B.S. in art. Chuck Close, he recalls, was a rare "positive source." "The second event was the experience of *being* an artist, not just *becoming* an artist." The third, which took place in '75, functions as a complex symbol. "I saw a dog fall out of a window. Maybe it was thrown out."

# GARY STEPHAN:
# THE BRIEF AGAINST MATISSE

Gary Stephan, now in his late 30s: steel-rimmed glasses, incipient embonpoint, well-cut tweeds — no longer the appurtenances of a stylish boy, but those of a stylish man. Not gratuitous. Art, for Stephan, represents — at least in part — a conquest of good taste paralleling the adult conquest of infantile neuroses. Yet all this is ambivalent — since his sharp sense of taste is a quality to be subjugated. Stephan, like us all, broods on the unraveling complex formations occasioned by family and upbringing; ambivalence animates his recollections, say, of a Levittown adolescence.

— "What were the '70s about," I ask, "and how did you fit in?"

— "The strengths and freedoms of the '50s led to the narcissism and license of the '60s. This led to a search for structure, for a logic behind the activity on which to base the activity. This belief in logic led to 'flat geometrics.' Soon we saw that flat geometrics were rigid, not optically responsive. Ultimately, they failed, and in failing, we were forced to turn back to appearances."

Stephan trained as an industrial designer at Pratt Institute, but went off instead to San Francisco before completing the degree. He recalls the Pratt formation as essentially Moholy-Nagy and New Bauhaus, very *Vision in Motion.* It was, after all, the prevailing art school mode persisting from the '50s.

Stephan arrived in the Bay Area of the '60s, a period when artists, especially California artists, were trying to draft a brief against Matisse. Ironically, official California painting today is dominated by the Matisse spinoff — a mode instigated by Richard Diebenkorn.

What this case signified for Stephan, at the time, was the frustration felt by painters identified as provincial and who partly believed the tag whether they wanted to or not — "provincials," then, both angry and envious of a formalism, the formalist criticism of a formalist clique, that excluded them: Greenberg and Co. All this is water under the bridge by now, and formalist criticism in the '80s — late semiological analysis, say — by no stretch of the imagination alludes to Greenberg any longer. Still, San Francisco as a place, as locale, continues to encode, for Stephan, the consequences of "narcissism" and "license," as he called it.

Stephan's departure for San Francisco has all the marks of a sudden impulse, an impetuous fugue. He enrolled at the San Francisco Art Institute, finding himself in the same class with John Duff, James Reineking — strong sculptors — and the painter of masks, Michael Tetherow. Bruce Nauman regularly dropped in on yet another classmate whom they then regarded as their most brilliant peer, a certain Randy Hardy.

Stephan's psychoanalytical bent was sparked by a set of early paintings about which he now feels a strong attraction/repulsion. We see them today as startling out-of-synch premonitions of the current New Figuration, Neo-Primitivism, Maximalism, what have you. Beyond this happenstance, his present masterful painterly abstractions on Christian themes or derived from El Greco's compositions (Plate 18) are internally linked in his mind "to the mad phophecy of these crazy student pictures."

Stephan showed me some black and white glossies of these early works — wide, hilly vistas, 10 feet across, depopulated and schematic, chimneyed houses with smoke coming out, all rendered in a dumb sleazy way, more Walter Lantz than Walt Disney. These large overlapped landscaped passages painted on masonite date to about 1964-66 and are punctuated with legible — if un-understandable — shapes, an odd iconography.

— "That 'odd shape'," Stephan says, "means roses that are chipped and wrapped in aluminum foil and then put in an ashtray." See what I mean, legible but un-understandable. Often enough, queer animals bear a resemblance to toilets or chemical alembics. Stephan vaguely recalls these "anthropomorphic toilets" (dumb cartoon flowers stuck in their bowls) as a distant model for the thin-armed "torso-shapes" that emerged in his paintings around 1980.

These quirky pictures were shown at the San Francisco Art Institute as part of his senior project. So unacceptable were they to the professoriate that he was obliged to take them down; only after protracted insistence that they were, after all, really meant to be "paintings" was he permitted to put them up again. (Well, a history of affront and contretemps attaches itself to the "M.F.A. Show" or the "Senior Crit," so ubiquitous has the degree become, that practically every artist nurtures a favorite horror story.)

On showing these works, if even in student circumstances, the question was at last asked of the artist, "Have you ever heard of Carl Jung?" Its reply led him to analysis and the belief in the insights available to painting that analysis provided.

These paintings are now stacked inaccessibly in the New York studio, but Stephan can get to at least one of them — *The Magic Motor Boat* (Plate 19), a scratched water view, a few scattered islands forming a narrow channel through which a motorboat passes, some lightning, a deformed horizontal shape.

— "The odd shape 'objectifies' the painting more clearly," Stephan says. Nailheads protrude from the scratched surface; again "greater objectification." Like a stiff altar cloth, the picture is hung in relief from an old bent curtain rod. "The curtain rod indicates the use of rejected cultural artifacts as an art strategy."

On leaving the cloister of the San Francisco Art Institute — its local quasi-paranoid tradition of the Transcendent Morality of Pure Painting still intact — Stephan returned to New York.

Of course, he had been picking up real-life clues all along, School of Hard Knocks — funnier perhaps in the retelling. It was 1968, and Kynaston McShine was installing the Smith, Irwin, Davis show at the Jewish Museum, back when the Museum had defocused from ethnicity for awhile. (Alan Solomon's first surveys of Johns and Rauschenberg are the now legendary inaugurations of this period.) Stephan and Neil Jenney worked on the installation. Jenney had heard Germano Celant's Art Povera book "chatted-up" in McShine's office, and knew that the future was in "The Conceptual" — dirt, plants, wrapped earth — you remember. So they made some Art Povera — I was calling it Postminimalism — and Jenney called Richard Bellamy to come down to inspect the stuff, with an eye to inviting them into the Green Gallery, then a leading avant-garde locale. A moment before Bellamy's epiphany, Jenney showed up in

Stephan's studio to borrow some geranium plants for his installation. Stephan lent them; and Jenney gets taken by Bellamy — with no follow-up visit to Stephan's studio, "though my work used geraniums too!" "It is not who does it first," Jenney later explained by way of exculpation, "it is who gets it first to the store."

But, it was not all self-service. Later, at an art party, Jasper Johns had asked John Duff to become his studio assistant. Duff, having sufficient funds on which to get by, recommended his classmate. "Johns," totally neutral and taciturn, "is the absolutely perfect Freudian analyst."

— "How much do you want?" Johns asks.

— "$3.50 an hour," the salary he had been earning as a part-time art therapist in a clinic on the Upper West Side. Johns nods assent.

— "And when can you come?" Stephan indicates the hours and the days. Again the nod.

For nearly two years, 1969-70, Stephan worked for Johns during the period of the painting of the second version of the Buckminster Fuller Dymaxion Map, arguably Johns' least agreeable work. Stephan gets to luncheon daily with Johns and his intimate circle, this by way of having invented the studio chores — sorting the mail, feeding the cats, stretching the canvases, and reserving the table at old Signor Ballotto's restaurant on Houston Street.

Such recollections are keys to work shown at the now defunct David Whitney Gallery. Whitney, a close friend of Phillip Johnson and Jasper Johns, had once worked for Castelli uptown. During the colonization of SoHo, Whitney had opened a galery midway down in the then Terra Incognita - 4th Avenue and the low 20s — more or less diagonally across from the old Max's Kansas Ctiy, then a preferred watering hole for artists. True, not SoHo — but not upper-Madison Avenue either.

The gallery partially reflected the combined tastes of Whitney, Johnson, and Johns, and can now be seen to have had its moments — especially the sponsoring, say, of the early Duff/Stephan experimental work and the early Neil Jenney installations. These highpoints were countered by much late formalist expressionism then called "lyrical abstraction," an abstraction I obtusely insisted upon calling "fat field." Set upon the gray industrially carpeted floors and walls of the top floor of a small loft building, the mix was novel and anomalous. Whitney's "go-fer" was David White, who also had started out as a Leo Castelli assistant, and the warm secretary was Jeanne Blake, who first "gal-Friday'd" in the old Richard Feigen place just off Madison into which Bykert eventually moved.

Of all the artists who exhibited with David Whitney during the two years of the turn of the '70s, Stephan was shown the most frequently, three times; and at length, he closed the gallery. His work, then similar to Duff's, experimented with rubber, latex, plastic, and free-process — a kind of constructed painting strongly marked by an awareness of pictorialized modernist sculpture (Plate 20). His works, at the time, offer no particularly lovable resolutions, though they were, for Stephan, highly therapeutic.

— "Being back in New York, getting rid of all that weird California shit, I tried to make the paint structural, to make the paint itself into the ground I was painting on. I see them as plastic paint."

My description of these paintings makes them sound too reliant on an admittedly self-evident expressionist bias — "pictorialized sculpture" — suggesting, in this way, that the paintings were worked in the absence of a specific set of informational cues. Quite to the contrary, numerous formal concerns are noted within this group. For example, any pictorial element that carried color was meant to be a specific formal cue. Such colored shapes, for example, always occur as "negative shapes" — though they are afforded an unusual assertiveness since they hold the strongest positive colors. The areas of the plastic paintings that were executed in black and white (or which were piebald or parti-colored) conveyed a sense of field or "spread." These "striped" areas, loosely speaking, connected up with one another across the shape interruptions that held the primary one-colored shapes together. Such stripes did not always take horizontal configurations either; at times they occur as radiating patterns. Last, a kind of order was imposed through the sense of symmetry occasioned by the strongly inflected bounding sides.

But, as Stephan notes, "To match these figurative cues is to deprive it of other things." In short, the paintings of circa 1969-70 attempted (through the application of an exiguous argument) to embody Stephan's desire to "circumvent style." Style, at the time, capital 'S' Style, was torn between the East Coast hegemony (a mode Stephan was, in some measure, attempting to accommodate) and West Coast latitudinarian laissez-aller (the dopey private mythologies he had abandoned).

With noteworthy Jesuitical reasoning, Stephan reduced this antagonism (purging from it its geographical references) to an argument in which Manet is pitted against Cézanne; or in an even larger context, Culture versus Perception.

How can this argument be schematized? Manet, for Stephan, emerges as a student of Spanish painting, meaning for him that Manet viewed historical issues as stylistic questions. "These issues seen this way engage the artist's technical ability at reproducing the diverse, smooth, or rough brushwork of his chosen models. A skillful adoption then allows the artist to be judged as either good or bad; and attendant historical issues then become conventionalized to those called up by iconography or iconographic interpretation."

Duchamp, in substituting mentality for retinality, rerouted this tradition through the introduction of paradox and irony — a content whose recognizability has scant shelf life. While the "thing" remains, the painting I mean, paradox and irony rapidly grow invisible.

Making a rather broad and pointed argumentative leap, Stephan sees this aesthetic curve-ball as a form of latent suicide, a self-destruction with regard to meaning in contemporary high formalist painting (especially that of marked iconographic profile — such as is now being made by Julian Schnabel and David Salle with whom, for better or worse, Stephan's present career is inescapably linked).

The warping of culture is countered in Stephan's argument by "perception," that is to say, by questions concerning what it is that one is seeing. Such straight observations lead to instrumentalization, to the tool rather than to stylistic questions. For Stephan, Cézanne inaugurated this aspect of the modernist traditon (as we all believe anyway), not Manet. Much of what Stephan calls "the perceptual tradition," we call straight formalism. Stephan considers an artist like Elsworth Kelly, say, to be an exemplar. "Culture, leading to Duchamp, leads to the consequences of 'reading'; whereas, perception leads to the consequences of sensory data."

Culture, then, leads to irony, and as Stephan notes, "Irony is a way to do what one does without assuming responsibility for the action"; or stated even more bluntly, "Irony is visual cowardice."

Stephan admits there is no pure case, though this really may repesent an attempt to slip off the hook, for, granting his hyperbole, Stephan also believes that "the social question always goes away. It always evaporates. The only things that have any chance of survival are the things that *remain,* remain that is *neurally,* not *socially.* That is where Piero is critical, and that is where Manet's occasional optical struggle to organize the spatial picture leads to less than authoritative spatial discrepancies."

— "What about quality?" I interject, "or our predisposition toward classicism over the long haul? Doesn't the inescapable binding of the notion of quality and the notion of classicism amount to an almost mystical connection in our culture?"

— "I don't believe in the question of classicism at all. Quality, for me," is neural, and the neural avoids the shoals of quality. Paintings do something. Like cars, they do something. And this 'doing' is perceptual and neural. Paintings operate. Something happens. That is why painting is a radical activity. It finally is anti-culture. It is in here (he says, pointing to his temples). It is in the engrams. It is purely Darwinian. Our receptors are set up through genetic encoding. Trial and error only help or hinder the process."

— "No," I interject. "That is too positivist a model, too materialistic. It's got no naunces."

— "Look, genetic shuffling brings the syntax flat up against evolutionary limitations. We're on our own. Shuffling occurs within us and within the limits of the enterprise, especially within the limits of the tools used."

If from the vantage of the early '80s the wacky mythologies of the California work, or the degraded rubber and plastic pictures, are perhaps more congenial than the informational aspects also present in the latter group, then the body of work of the early '70s is perhaps the most displaced. In the measure that an epistemic argument for the arts of the turn of the '70s took hold, Stephan's paintings were thought to reflect this tendency, one that favored information. In the degree that Stephan rejected confusion for simplification, the marriage of incipient minimalism to a waxing sensibility became evident (Plate 21).

It is, of course, the classicizing element of Stephan's work that would carry him through this moment despite his disavowal. His first successes were appreciated by a circle of artists and connoisseurs deeply associated with epistemic abstraction. Surely the soundest advice Stephan received in the early '70s came from Richard Serra (advice tendered during the group show of new tendencies organized by Jay Belloli for the Contemporary Arts Museum in Houston, Texas, September-November, 1972): "In a work, never do anything more than once."

This minimalist maxim helps explain the reductivism of Stephan's work throughout the decade. All the consonances and dissonances of shape, scale, and color relationships separate out; relationships simplify, providing only sufficient compositional fodder for single organizations, when previously such material had served as units of far more complex, polyphonic structures.

Two elements, reduced dualisms, now suffice to justify the organization. Briefly, composition became that of a reductivist formal strategy: How does one scale or locate a figure? How do you validate a figure in its place? And how can the artist continue to reduce the number of pictorial elements from picture to picture? The answer to these questions at last led to the emblematic gray square that haunted minimalist painting during its sway, and eventually occasioned the sense of the utterly contrived and academic which sounded the death knell of the style.

Stephan bridles (as any artist would) at this formulation of minimalism's evolution. He would point instead to the "neural" validation of shape and form. Shapes were thus and so because they inescapably had to be that way, not out of any accommodationist tactic to a larger controlling *Zeitgeist.* For the public, though, Stephan's pictures at the time were seen as the chief representatives of a minimalist gradualism, the vague coloristic and tonal aspects of which continue to place his work outside the theoretical pale of the absolutist "heavies," a gradualism that continued to speak for pictorial qualities then seen as of dubious consequence. Among these properties were Stephan's "designerish" organization (from Bauhaus graphics, say) or a color connoting a broad range of evocative association. These were regarded as sentimental hangovers that fell outside the pervasive minimalist context. It is precisely in the degree that these works do fall outside the precinct that they are today able to be seen quite freshly.

Naturally, for Stephan, the whole issue of color and formal connotation is false. His view of color at the time was one of neurological absolutes: colors had to be (and are what they are) not because of a palette predicated on association, but because of a concentrated focus, a bearing down on the issues of pure abstraction. Then, if one speaks of a Mediterranean palette, or delicate ranges of *greige* or flesh, or refined hues as indications of a drained and tasteful imagery, such associations are made on the viewer's account, not the artist's.

All this nuanced color is countermanded, as it were, by essential geometrical configurations based on simple weight displacements — a circle placed against a triangle, a semicircle against a rectangle. Such compositions are rendered in an equally reduced palette, one that seeks to establish no more than figure-ground relationships despite the unswerving delicacy of the colors assigned to the figures and the ground. In this way, one may continue to make dubious bracketings between the pronounced sensibility minimalism of the '70s and Stephan. Figures like Brice Marden, or Ralph Humphrey, or Doug Ohlson, or Doug Sanderson, or Susanna Tanger all come to mind and all are, quite naturally, wrong in varying measure.

In truth, the minimalist graphic attachments of Stephan's work of the early '70s represented a return to painting itself, if only by fits and starts. Naturally, there had been the example of Ellsworth Kelly whose work — though reductive and itself even attached to the prevailing epistemic enthusiasm of the mid '60s — grew out of an abstraction grounded in distillations of landscape and architectural motifs. This abstraction as an essence of nature is, for the modernist, its most venerable form.

Still, for all that, the period is marked by a hostility towards an overt and unapologetic painting. Indeed, painting itself was taboo, and even as they painted, Stephan (and many others) regarded their work as akin to the analytic reduction of the clear informational structures of early minimalism. In short, they worked from a hobbled and apologetic stance. It was not until the middle of the decade that the need to rationalize and justify the very fact that one was painting faded. Work, as it were, stood on its own, freed from the highly referential systems imposed by epistemic abstraction.

The Garden Cycle, a group of ten approximately 44-inch square canvases, is characteristic of the shift. The paintings were shown at the Bykert Gallery in 1975. They may be understood as demonstrating elementary figure/ground relationships rendered ambivalent through spare illusionistic bends. This reading of convexity was occasioned by applications of denser pigment. At this moment, Stephan's characteristic impasto, a build of graduated crust along a slightly scored edge, becomes quite incisive; equally marked is his still somewhat apologetically adduced historicist iconography and associative color. The instances of powdery brickish colors were specific to Stephan's art book study of the program for Giotto's Padua Arena Chapel, the fresco cycle depicting scenes from The Life and Passion of Christ.

The very use of the word "cycle" in the Garden Cycle was triggered by this connection. Not only did fresco technique provide a range of chalky color relationships; it provided, as well, an isolated focus on curious architectural settings and motifs to be seen throughout the program. For example, the section depicting Anna and Joachim (the future parents of the Virgin) meeting at the Golden Gate — Giotto's imaginary Jerusalem — provided a set of architectonics which, when even further reduced, appear in Stephan's pictures as abstract figure/ground relationships emblematic of the distilled architectural motif.

Occasionally, the shapes are not modified at all; rather, they are "lifted," traced from the compositions reproduced in glossy picture books or popular guides to religious art. Thus the figure/grounds, on a certain level, may be read as excised turret or fortress dentellations. In certain ways, bold elements were meant to transfer the massive sack-like figures invented by Giotto suggesting, to Stephan, "the human plus its metaphor in architecture." All this *italianismo* comes to a head in 1975. Not only had Stephan visited Venice, California, in that year, he also made his first trip to the "real" Venice, sojourns that had the effect of confirming the incipient Italian connections — associations, like particles of metal, that came to adhere to the magnet of minimalism.

So excited was Stephan at this discovery, he communicated it to Brian Hunt, then beginning his own career in California. Hunt, excited by Stephan's discussion, literally built a model of the Golden Gate at which Anna met Joachim. Surely for a Californian, erstwhile or otherwise, the coincidence of "Golden Gate" and Golden Gate would be very striking, as is the parallel Venice/Venice equally arresting.

Thus the Garden Cycle is a culminating effort for Stephan. Each element encodes highly conscious choices for the artist. Its shape, for example, purposefully equivocates on the format issue, being neither vertical nor horizontal. The implication of the square format was first guessed at during the Venice summer. A vertical format, Stephan suddenly grasped, invoked a mimetic "other person," a peer and equal out there in the world. By contrast, the horizontal format came to conjure the spectator who looks out over the terrain. In this sense of inspection, the "voyeur" (too erotic a notion) or the dutiful Christian (associated with a reconstruction of the lost earthly paradise) was invoked. But the square itself waffles.

Stephan's perceptions were, of course, part of the growing awareness of the pronounced iconographic classicism that attached itself to minimalist logic — Ellsworth Kelly's spare balances and swells born of an acute appreciation of classical epigraphy and letter forms, Bochner's *sinopia* and dry pigment fresco technique, Rockburne's golden sections and her own version of *Maestà* imagery, Marden's Fra Roberto Caracciolo of Lecce, Mangold's Albertian church facades, Novros' *predelle,* etc.

Behind much of this Pre-Raphaelite overload is to be found the neglected figure of Gordon Hart who, like many before, had been stimulated by Cennino and the recipes necessary to early Renaissance painting — grinding one's own pigment, setting down one's own perfect gesso ground, burnishing the hand-set gold leaf, rediscovering the lost techniques of egg tempera, studying the magical ratios of polyptychly subdivided surface, and on and on. Hart's career, though checkered, was pervasive in the '70s, a prestige heightened through his friendship with Barnett Newman. "While he didn't get to run with the ball," notes Stephan, "he was the one who made the connection."

The use of gold leaf in Lynda Benglis' sculpture, for example, may be traceable to a residual connection to Gordon Hart. On her arrival in New York from New Orleans in the late '60s, Benglis, greatly associated with him, a point of some interest insofar as it was across Hart that Benglis first met Klaus Kertess, the director of the Bykert Gallery, with whom she entered a friendship that spanned the decade.

Briefly then, at the very least, Sensibility Minimalism of the '70s may be viewed as The Bykert Problem, insofar as Kertess' highly sensitized taste brought his gallery to the pivotal place it enjoyed at the time. We have already traced Stephan to the closing of the David Whitney space on Park Avenue South; indeed, it was on the closing of that gallery in 1972 that Stephan joined the Bykert stable. Before 1972, Kertess had been installed in a small office-like space on West 57th Street. Stephan remembers visiting him there, sitting at his makeshift cardtable desk with these "self-conscious mauve boxes posed on the floor," wondering, "What is all this shit?" From there, Kertess moved up to 81st Street, settling into the old Richard Feigen quarters and remaining there until the gallery closed in 1976. The move uptown was made during the summer of 1968. Stephan, by way of supplementing his still uneven income, worked as library cataloguer for Feigen in anticipation of the latter's installation in an elegantly renovated 79th Street townhouse. Gallery chores were shared with Frederica Hunter, who would shortly open the Texas Gallery, still one of the better galleries devoted to reductivist minimalist taste. Need one add, by way of noting the extraordinary network of apprenticeships, that Kertess' last gallery assistant was Mary Boone, whose early taste as a dealer was initially formed by the gallery and by her strong personal relationship with Kertess, Benglis, and Stephan.

"The thing about Klaus that was so striking to artists," Stephan notes, "was this unnerving ability just to look and look at a work, for a long period of time, staying with it longer, even, than the picture itself worked. Klaus, in a way, could watch a picture end. I think that is why," Stephan continues, "so much of the art shown at Bykert was tinged with melancholy. They were seen through."

Kertess' most profound friendship from among his roster of artists was with the young Brice Marden. In a certain way, Kertess can almost be said to have been under his emprise. The intensity of the connection inescapably marked the gallery's orientation. Stephan's belief in the neural engrammatic character of painting, its sheer essential quiddity as deriving from Cézanne (not Manet) was shared by Marden.

This staying with the work, Kertess' way, Bykert's way, Marden's way, may be taken to be yet another facet of the strong rejection of Greenberg's modernism felt at the time. Greenberg had postulated the original sanctions of the modernist dialectic as stemming from Manet. An essential character of modernist painting at the moment of its failure in the mid-'70s was its presumed instantaneity of perception. Painting, by way of differentiating it utterly from sculpture, had to be accessible within a single glance. This way of seeing equally was countered by a then contemporary reassessment of the work of Reinhardt, in whose painting glossy instantaneity was replaced by time-consuming, long looking. Thus it disloged painting from its role merely as a screen of delectation to become, instead, one of meditation, as Reinhardt would have it. Need one note, then, that among the canonic field painters as determined by Greenberg — Newman, Rothko, Louis — Reinhardt's name does not appear.

Indeed, Stephan's first memory of a significant art experience is intimately allied with Reinhardt, but much before this date, when he was a callow mid-Long Island teenager who had been taken to see a Reinhardt painting, part of an exhibition organized by Vincent Price for the New York Coliseum; and like the rest of the teen-hoods he was with, Stephan merely laughed as he was paraded through the gallery. An astute teacher knowing him better than he knew himself chastened him and sat him down before the

Reinhardt: "You just sit here and look hard," she said, "and you will see something you have never seen before." After a while this fuckin' black cross jumped out at me from the canvas," Stephan continues. "I was so excited. When I got home, I had to make black on black pictures; and I had just seen Jackson Pollock on a talk show, too," which means this memory dates to 1956 or earlier. "The point was to make fun of his 'dripping,' because they put him on with this jerk — WOW — who made a pair of wings to fly with. I put Pollock and Reinhardt together, mixing gravel and paint and throwing it in the form of a cross on my father's shirt cardboards. Then I made these crosses mixing Karo and chocolate syrup on the cardboard. They were slow gravity pictures and had to be turned once a week or else the stuff would fall off onto the floor. And my parents turned them faithfully once a week without complaining. They never really impeded anything that had to do with art for me."

But instantaneity is but one factor of the modernist problem — its essential question addresses the illusion of flatness. Stephan counters that "painting's true space is all the space it is" — an affirmation of the artist's essentially existential and phenomenological attachments.

These, of course, led to the rejection of the progressive or evolutionary character implicit to Greenberg's notion of modernism, the notion of art as a result of a dialectical process. Of course, to reduce Greenberg's historical machinery, to regard the procession toward literal flatness from Manet through the formalist painting of the '60s as merely a progress from illusionistic space to flatness, is to deprive it of its paramount mystical feature, its tacit acceptance of the condition of "quality," a condition that eludes all mechanistic use of history. Even for Greenberg, quality is a transcendent inspiriting force fully palpable to the true connoisseur, but invisible to the laity. It is not so much that two pictures may present concordances of flatness, but that despite this concord, one still may note that one space is "better" than the other or "purer" or "finer" or "deeper."

For Stephan, painting is constantly rejective of notions of transcendence, as indifferent to academic norms as to cultural desiderata. Painters are involved (as are all sensitive artists) in the instrumentalization of their work, another way of saying that they keep their options open; but this in itself may yet prove to be simply an updated reprise of the Marxist dialectic necessary to Greenberg, whether one likes it or not. For Stephan, "artists simply find ways of doing things," admitting all the while "that their work is limited by their personality."

Granting this, they ought not (as the expressed intention of painting) engage in "personal production," ought not make "ego-directed choices," ought not "agonize over how the work will 'look' or worry about 'style'." Instead, the enterprise of painting has a kind of inadvertency attached to it. Things all happen "because there is no way for them not to happen." Through sheer repetition of the activity of painting, "the unbidden subtext will emerge anyway, those formal and iconographic considerations through which we recognize an intelligent work."

In a curious way, Stephan seems to be talking about statistical occurrences as a kind of demonstration of formal and iconographic content. He rejects seeing the intention of painting as merely the forcing of their appearance out into the open. When this happens, he speaks of a manifestly "distressingly subtexted picture." Thus painting for Stephan may become a kind of diffident exercise, a pragmatic activity engaged in without an end in view. As activity, it is all filler material. Formal values and iconography happen anyway, so why worry about them?

But all this has led me from The Bykert Problem. Kertess himself came to outgrow, if not resent, his "helpmate" role — the myriad insistent details of "just running" a gallery were, in the end, inimical to his ambition of being a writer. And he closed. At the same time, his attachment to his artists assumed the demands of their own destinies - in Marden's case, to the considerable glamour of the Pace Gallery and its powerful support system. For others, the situation they found themselves in did not prove so easy. The psychological sustenance of just being able to say that one is represented by this or that gallery is an immense aid; this whether or not one's work sells.

The closing the the Bykert Gallery was coincidental with the incipient, turbulent expressionist resurgence, one that I have come to refer to as Maximalism, much of it stimulated out of sheer despair with so long a diet of Reductivist Minimalism. The bridge to this mode, so characteristic of the present moment, led to Mary Boone who, sloughing off Bykert, continued to enlarge her own taste, at length to support the more radical seeming palimpsests of Schnabel and Salle and other Maximalist Neo-Expressionists.

Perhaps, by way of conclusion, a word separating Stephan from Schnabel and Salle is in order — especially after having forged so specious a link between them. The fundamental difference between Stephan and the others is primarily one of generational attitude. Stephan's first convincing style comes to fruition within a Minimalist frame of reference, theirs in a Postminimalist. Thus, there must be between them — on an artistic level — a profound antagonism. Stephan, like the other artists who matured when he did, can easily, naturally embrace the history of art. He can feel his work, in some sense, to be in continuity with it, can feel his work, in some natural way, to continue a self-aware tradition of high art, and can view this continuity as normal, associated with the grand epochs and great figures of the textbooks. In short, he can feel at ease in drawing connections between his work and the artists and heroic "isms" of early modernism - especially Mondrian, Malevich, Suprematism, de Stijl — and so on.

By contrast, generally speaking, this cultural baggage provides a kind of embarrassment, in both senses, of riches and emotional disturbance, to the younger painters. They feel, in some measure, ill at ease and a certain bad faith at the seeming pretensions of such adduced associations. Of course, words like "bad faith" or "embarrassment" say nothing of quality. Both Salle and Schnabel, not to mention other generational compeers, are quite simply superb painters. My words are meant to speak to a certain cast of mind, a position of awkwardness vis-à-vis a cultural history that, after all, has the effect of generating compensatory systems of ratiocination, a need to justify their arts and actions in ways that are far less than convincing than the art itself. By contrast, the sovereign security of their painting is never in doubt.

In further contrast, Stephan does not believe in a generalized imagery of fractured and dispersed subject matter, but of highly specific and reduced information. His painting is not bidden by the democratizing visual gambits utlimately ratified in the Combine Paintings of Rauschenberg. Stephan celebrates his sense of profound continuity with tradition, his emotional ease with the past. "I want to reassert a sense of Mission. They," meaning the Postminimalists and Maximalists, "need to cover their sense of shame about tradition through visual sarcasm. I wish I could feel what they feel. I wish I, too, could feel free of history. But, I'm not free of it, so I have to take it all in and still hope to make 'normal objects' — we call them paintings — objects that sit comfortably with the facts of themselves."

# ERIC FISCHL: FREUD LIVES

*10 May 1981*

A complex and rich panel discussion at S.V.A., organized by Joe La Placa and Randy Black with Tim Rollins, Diego Cortez, Jenny Holtzer, Barbara Kruger, Allan Moore, Julian Schnabel, Ingrid Sischy, and Carol Squiers. Rollins and Cortez the most impressive. Rollins' essential point was that artists must hang on to controlling the meaning of their work for as long as possible. If artists cannot state what the meaning of their work is, then they have that much more rapidly lost control over it; for, if they cannot say what this meaning is, someone else surely will. Artists must exercise vigilance with regard to this statement so as to forestall meaning from inevitably being misrepresented in weak, sentimental, or just plain wrong terms.

In a capitalist bourgeois world, painting is inevitably reduced to its decorative common denominator which, after all, just may not be so terrible as all that. The forestalling of this inevitability can only be achieved through the artist's control over meaning.

This startling and fresh idea was seconded by Barbara Kruger whose artful critical lingo was as striking as her unsentimental way of handling a barrage of clichés about art offered up from the floor by the poet-cum-critic René Ricard.

As striking as Tim's insight was, so too was Diego Cortez's burning anger. Cortez spoke after Allan Moore personified the effects of a lingering flower-childhood, Jenny Holtzer that of maternal openness, and Carol Squiers that of hyper-personalistic professionalism: "I will try to be at least as interesting as the last three speakers," Cortez said facetiously.

Cortez has grown more punkishly anti-liberal in the degree that he has come to be perceived as a certain kind of New Wave artist and animating force. Feeling ripped-off, pragmatics suddenly concern him. Cortez is interested in "worth efficiency." After the immense social success of the P.S. 1 New Wave show, which he organized (and which will travel), and angry at the "exaggerated factionalism" that the very success of the show generated, and bored by "the exodus of artists into the music scene," he now rejects out of hand: Robert Longo, *Semio-Text(e)*, *Soho News*, Pattern Painters, most "groups," Metro Pictures, Stefanotti, "Energism," alternative spaces, the Times Square Show, etc., etc. All in all, the list was hilarious in a *dernier cri* in-group sort of way, but at length despair and frustration sparked Cortez and he walked off the panel.

Something of his anger was rooted in a thwarted awareness of all the mileage accrued to the rest with no fallout for him. New Wave as a style is merchandisable. And yet, at its inception (as if one could ever locate such an instant), New Wave was militantly anarchic and anti-careerist. But that was a long time ago. Now, I suppose, after there's been so much cashing in, Cortez wonders where his share went.

Tellingly, Cortez worried at the most exposed nerve of the evening. An unlistening crowd responded negatively to Julian Schnabel's remarks. After all, though his appears to be the most violently successful and sudden emergence, it's really not been roses all the way (It should be noted that the broken crockery of Schnabel's so-called "plate paintings" may take heightened cue from the grimy years he spent as a dishwasher in crummy SoHo eateries.) "How many of you," Cortez asked of the resistant crowd, "are just plain jealous?" For sure, many; for sure, most of the hostility is rooted in this emotion.

Then Cortez scatter-shot. If in his poverty he has found a way of self-presentation with great flair, then why deny this possibility to Tim Rollins? Cortez should have seen that Rollins' red jumpsuit was just common duck worn by a kid with an innate sense of style. Rollins, authentically stylish and authentically activist, spoke of Group Material and P.A.D. (Political Art Documentation/Distribution), noting that the former was not a "vanity co-op" on 13th Street but a form of "organic democracy" that seeks neither to "fetichize Hero or Collective."

But Cortez is sharply cynical and viewed the 13th Street effort along with, say, Fashion Moda or Co-Lab as yet another form of "racist colonization," despite all claims to the contrary. Joe Lewis, a dreadlocked black artist from Fashion Moda, questioned from the floor about the possibility of "New Wave Quality." (Imagine, the sanctions of Greenberg's aesthetics of privilege called up in the most unlikely place!) Then, he offered a deeply convincing definition of that same quality as being the evolution of "something you don't know to something you really don't know."

*3 June 1981*

A compelling deception inhabits Eric Fischl's suburban world. One is thrown off track by his skillful faux-naif execution and a skewed expressionism intensifies a cunning iconography exactly suited — image to image — to a classical Freudian skin-flick.

A so-so loft on Reade Street, up five flights and an occasional chair covered with cat hair. Several large canvases still being worked on. The first, *Bad Boy* (Plate 22), represents a boy-child rigid at the sight of his mother's(?) large, dark genitals. The scene is seen from behind and we gaze as he gazes upon this Oedipal trauma. His right hand is placed behind him, entering a purse (penetration/violation?), stealing money from mom's store of bucks, if I may phrase it this way, Freud lives. Two forbidden acts: theft/the view of the genitals. The hand slipping into the purse signifies the Oedipal fornication that psychologically precedes all others.

I am of two minds about a second large canvas which may be, perhaps, even finer were it not for a certain willed awkwardness in execution that is hard to get by. This canvas, *First Sex,* is even more cinematographically framed than the first. In it, to the left, sprawls an immense earth mother (Gaia?), legs akimbo, tantalizing and receiving an erect little boy (Gaia's, copulation with her son Uranus from which union sprung the father of the Olympian gods, Cronus?). Toward the left a genital-cupping clothed celebrant may signify carnal knowledge and as such, may be the boy's father (Cronus, father of Zeus?); if so, scale-wise he is either too small or the boy is too large. In the far background scumbled into the grayish paint, a campfire dance of Bacchic initiates.

The most impressive painting, *Time for Bed* (Plate 23), depicts suburban frenzy. A drunken mother (?) madly dancing on a glass-top coffee table spills a cocktail across the polyester-like summer togs of a maimed father (?). He has lost his left arm, a slack sleeve tucked into his pocket. The lost appendage, for sure, signifies castration. This lost arm invokes, say Kirchner's handless wounded soldiers and still other examples of the Freudian awareness of much German Expressionist painting.

Risking rebuff, I jump ahead by two steps: "When did you discover Beckmann?" I ask, the stylistic connections between Fischl and Beckmann amounting almost to homage. "*Departure,* that big painting, taught me how to look at painting again."

To the left of this scene of suburban carnage are two children, a skinny boy-child in Superman p.j.'s (the emblem shield "S" is inscribed backwards). Is Fischl a touch dyslectic? It seems so. "I know it takes me a very long time to read things through." Superboy's ungainly kid sister is seen from the rear, her hair towel-turbaned for quick drying. Her awkward dislocation echoes Mother's expertly rendered high-heeled shoes.

Fischl's uncanny eye builds from detail to detail in an unnerving way. "I work with single events; I just go with images. I had to get over my own sense of embarrassment." On the patio seen through the "walk-through windows," the neighbors (?) dance wildly, a reprise of the dancing figures in the beach firelight. To the left of this scene, a large black idol — middle-class primitive art — asserting through its presence the whole myth of black sexuality as counterfoil to Father's impotence.

Contrary to my snotty expectation, Fischl is not a Long Island Jew, but a cautious, taciturn 33-year-old Protestant who hails from Port Washington. On the death of his mother (the drink-spilling lady?), he moved from Long Island to Phoenix with his father (the man with the injured arm?) who remarries there. In time, Fischl goes to Cal Arts. Indifferent to that school's heavy emphasis on the West Coast conceptual tradition, he stayed close to the traditions of refined minimalizing abstraction, studying mostly with Paul Brach and Allen Hacklin. Still, the paintings reveal West Coast fascination with camera technique, parallax distortion, Mannerist scale differential, pan shots, and the like.

The actual application of paint itself is rooted in a loose gray-scale painting, a grisaille permeated with chromatic hue — but in itself not naturally the color of a native colorist *au but des ongles.* But it is not arbitrary color — if there ever can be such a thing. Indeed, another work, *Birth of Love,* transforms this grayness to a kind of glaucus chlorinated light, almost yellow-green rather than sea-green. At poolside, a naked boy-child massages the thigh of his naked mother — Gaia again? On the water's surface float two great vinyl beach balls — by Fischl's own account representing the testicles of Cronus, those hacked away by his son Zeus. Remember, these mutilated chunks of the Father's body when thrown into the sea churned up the waters occasioning the Birth of Venus, Venus Anadyomene, Venus born of the froth or the foam.

At Cal Arts, to judge from slides, Fischl was a competent abstract painter of minimal orientation with just the barest suggestion of a personalized iconography breaking through a hard reductivist shell. The abstract image is that of "a house in which the Blindman lived" (Oedipus? Oedipus blind equals Oedipus castrated). "My abstract paintings were brought up to a certain level of competence. They had a certain presence. Then I reduced the image to that of a house, an object. The house I saw as a shield, a protection. From here I could go inside the house and abstract the table and chairs. But such objects were insufficiently ambiguous. So from here I began to deal with the figure, if even crude or primitive, and I began to work the glassine figure drawings one on top of the other. From here came the Blindman."

About five years ago, this substitutive process, suburban imagery shot through with classical iconography, freed Fischl from continuing on in a tradition of merely tasteful abstract painting. "I wanted to tell the story of the picture but without recourse to codified and analytical literature." Indeed, Fischl has not studied either analytical literature itself — he was not a "psych" major, nor has he been on the couch — nor is he particularly expert in classical legend. Yet his images never fail. He's not been analyzed, is not a great reader; yet this self-absorbed "conversion process" outlined in the preceding paragraph led him back to figuration.

# STRATEGIES WORTH PONDERING: BOCHNER, SHAPIRO, LeWITT

*8 January 1978*

Mel Bochner's show opened yesterday — in the front rooms of the gallery there were magnificent wall works (Plate 24), in the rear space, a range of beautiful framed pastels. Despite all this, my first (and, as will be seen wrong) take: sensibility recidivism. Within the large formats occasioned by Bochner's 3-, 4-, and 5-sided shape permutations there was the additional clutter of elaborate "pure feeling" arrangments, Suprematist-like constellations at their most decorative — here and there cropped and near-accidentally placed in the larger formats (or so I thought at first). Hence, the triangle, square, and pentagon-linked master "templates" in Bochner's work were countered by an internal design of the kind that the Suprematists themselves had applied as ornament to classical porcelain shapes codified throughout the 19th century, utopian signifiers pointing up an almost headlong cultural collision. Here too (I thought) the exchanges that took place between Bochner's internal patterns and the external holding "templates" of the master perimeters were very beautiful and very hollow in the same way unless, of course, Bochner's Malevich-like sense of ornament was in fact cropped and copied from specific Suprematist organization; but that would reveal an epistemic irony natural, say, to a Benni Efrat, but anathematic to Bochner's high-minded seriousness. Were this even a remote possibility, then it would serve as further testimony to the admitted thralldom within which great epistemic abstraction labors vis-à-vis the Cronus forebears of Suprematism.

*15 January 1978*

Back at the Bochner show and finally saw that the variations that transpire within the larger 3, 4, 5 figures, for all their seemingly errant play, are generated epistemically. But they function in a scale at one remove down from that of the master edges. There is no intermediate group of shapes; they would have read as too close in size to that of the master shapes. Thus, the internal figures are quite small when compared to the overall figure. The smaller shapes negotiate a tripartite crew of linear kinsmen: oblongs, bars, and lines — the last I construe as slots since they incorporate the ground of the wall inducing an occasional spatial illusion. Bochner's patterns, when scaled down from the initial edges of the master shapes, constitute a concise set of linear figures — oblongs, bars, and lines.

They recall the wooden "lines" that occur on the walls, say, of Le Va's spatial projections — they too are epistemically generated — or the lines, tears, and folds of Neustein, say; or LeWitt or Rockburne for that matter. And for the same reason, epistemic linkage.

But Bochner's use of color is utterly non-epistemic, arbitrary, though to use a word like "arbitrary" sounds negative. I don't think so, at least in Bochner's case, as I consider him a colorist to his very fingertips, though no theoretical premises govern his selection of colors except, of course, that they occur as a result of felt necessities. Such necessities are in the end, only felt, and ultimately justified (perhaps) through the esteem that history may accord the work, but not through the application of any demonstratable system. The same may be said of Rockburne's color in her more recent works — she would have it that the "weight" of a certain shape demands a correspondingly "weighted" color. But to invoke this argument, though it never can be true in the sense that an episteme is true, invokes the specter of the early school of modern abstraction that taught the essential lesson — the congruency of form and content.

Still thorny is a tellingly reasonable comparison with Frank Stella's relief paintings of c. 1972 and thereafter. Stella's work at the time joined incisively formatted shapes of a clipped and snappy stylishness, internal hierarchic scale comparisons, all manner of tipped plane and substance, and an unworried color. Such good-looking Neo-Deco design would seem to have provided Bochner with the model of work by a celebrated artist at but only a generation's remove — had not Bochner specifically rejected Stella early on — in a review in *Arts Magazine,* May 1966 — when Stella was widely taken to be the Compleat Young American Master. Bochner rejected Stella as much for the latter's felicitous sensibility as for a concomitant absence of epistemic base — unless, of course, a season to season shift in strategy may in itself be a kind of formalist episteme.

Still, there is the matter of Stella's early and great "pinstripe" paintings — dating to 1959; they pointed to Malevich as Bochner's continue to. But Malevich, as we now know him, is largely an art-historical reconstruction that follows hard upon, not precedes, Stella's "pinstripes": or even the stretcher-generated shaped canvases. The reconstruction and wide rehabilitation of Malevich specifically and Suprematism generally coincides to a moment in Stella's career when the latter's avowed commitment to Matisse was inescapably self-evident. What may be interesting in this adjudication is that Matisse, who lurks behind Stella in all his moves — even the earliest (the white lines of the "pinstripes," by the artist's admission, allude to the blank canvas lines of Matisse's *Red Studio of 1911)* — may be said to animate certain of Bochner's recent moves too, as in fact it always has — the "Belief System" part of Bochner, the part that epistemology doesn't and can't explain. True, Bochner eschewed Stella but Matisse was never off limits — and the Matisse of the Barnes Foundation murals (1931) through the cut-outs of the 1950s provides Bochner with a revered model of sensibility-linked painting — quite in the same way that it equipped and enriched Stella's work.

Thus, if all the preceding is true, despite Bochner's rejection of Stella there remains some narrow sector of overlap between them, some communal patch, since it was Stella, above all the young American masters, who unwaveringly pointed to Matisse from the outset, semaphored the message bright and clear — before Bochner and the affiliates of Epistemic Abstraction's lean and pared-down look took up the signal flags — and indeed of late some have with especial alacrity. In Bochner's case the epistemic enterprise still obtains, but in the work of others now keeping or parting company, mutatis mutandis, Bruce Boice, say, or Dorothea Rockburne, they have all but abandoned the claim. Rockburne minces no words: over Christmas eggnog she said, "It was never about the system — it always was about sensibility."

All this preoccupation with Bochner put me in mind of dear dead days — when Bochner, not quite driven from the fold, was nevertheless excluded from the clubby set of godfathers of the larger Conceptual scene. At the top of the heap, the *capo dei capi,* so to speak, were Carl Andre and/or Sol LeWitt — followed by lesser *luogotinente,* Joseph Kosuth, Dennis Oppenheim, and groping somewhere still further below, Robert Barry and Lawrence Weiner — and they in turn condescended to the Art & Language types.

Those were tinctured days; somehow Bochner was getting mileage without queuing up in proper pecking order; nor would he signal in proper body language — you know, one's head slightly lower than that of the cluck above. How chafing the derision and now that he continues to garner esteem without ever having worshipped at the altars of LeWitt or Andre (and, mind you, I think them entirely worthy of the tribute) somehow further aggravates the scene. (Not to "worship at the altar" is a far cry from not esteeming. Bochner has always valued LeWitt as an artist.)

*1 February 1978*

Yesterday, Joel Shapiro — 36, teddy-bear-like (fighting weight), black thinning hair about a Claudian Charlie Brown face, white painter's ducks on a cold day. He is always fearful that he is not quite articulating sharply enough — though he talks clearly and securely about art. But he is unpolemical in his views, so that from sheer pleasure of company one neglects to write down what he says and then, pressures of the day being what they are, one forgets his exact phrases.

Shapiro's most conspicuous feature as an artist has been the willful bending of an epistemic art to seemingly aberrant personalistic issues. His timing was funny, albeit deft and light-footed — sort of a sprocket-jumped film strip. At the moments of tight epistemology he drew back in the name of stray obsessions; conversely, as the orgy of self-exposure, the loquacious need to expose one's most telling peculiarity began to cast a fatal appeal throughout the art world, so did Shapiro once more affirm solidarity with the coolest strategies of the epistemologists.

Shapiro's quirky feints were worked out, say, in his obdurate commitment to an obsessionally small scale — though there are models: Duchamp, Johns, and Tuttle — while the declarative public constructivism of Serra and Andre set the sculptural paradigm. Shapiro generously denies nothing to them. His capricious need to expose a privatistic symbology within an art that was, before all, one of stern formalist purpose was perversely captivating — those little Monopoly houses, doll-scale chairs, toy horses (Plate 25). But a pure investigation of a system — LeWitt's mostly — leads Shapiro to an enervated lassitude largely because for Shapiro, LeWitt is unscarred by the sculptor's brand. Hence, LeWitt's work, for all its manifest significance, is, for Shapiro, in some integral way, not sculpture at all. And I agree with him.

By contrast, LeWitt's great investigation of a whole system really amounted to the imposition of an alternative to formalist abstract values, values that still cling to Shapiro's conception of sculpture. For all his peculiarities of scale, all his tentative anxiety, Shapiro is psychologically fixed as a sculptor. He has made a primordial commitment to the species — yet, ever anguished, he continues to work within a framework of tantalizing ambivalence.

Thinking back, I try to mentally reconstruct his career.

67-68? the eccentric pictorial sculptures, the hairy canvases that form part of the larger concern with the Abstract-Expressionist revival that marks the late Sixties.

69? the shelves — the idiom would later emerge importantly as the eccentric supports for aberrant "houses."

70? the epistemic comparisons of the weights of primary raw materiel — here the allegiance to Serra and Andre is most explicit — so and so much metal compared to a like weight of stone.

In a parallel mode, the 70-71? comparisons between hand work and machine tooled, an episteme marking a tactile scale between coarse and fine — elegantly floor-bound proposals that assist in recalling what the old Paula Cooper Gallery was then all about, up three stories on grungy Prince Street. Part of the reason that one waxes

nostalgic is that one would never have supposed that the present epistemic prophylaxis would have been so complete.
71-72? the little trestle bridge, the whittled balsa boat, the pathetic little bird, the first "Monopoly" house, iconography of high-profile dream analysis presented in the most inaffective way except, of course, for the sheerly profligate drama of presenting tiny pieces on huge floors - an echo of the floor-adhesived masking-tape theorems of Bochner, then being shown in the unexpected emptiness of big gallery space. Brancusi is also a prototype — all those essentially Symbolist disembodied heads presented as the sheer stony weight and compression of unapologetic lithic molecules.
72-73? the little house placed on eccentric spaced shelves, elongated or cut, the toy man strutting in the direction of the cardinal points like some robotic Bauhaus logo and from 1974 on, the house itself studied and restudied, opened up, sectioned and bisected, a studious commitment to grasp the ultimately Cubist-connected formal content of this little temple form that would, in time, lead to a greater and greater analytical statement, the most recent works in fact. When seen from above, say, bird's eye, the sculpture is transformed into pure shape relations, the floor playing ground to the sculpture's figures. The imagery now annexes Malevich — and, in this, is again following Bochner's mode — yet when the new work is apprehended from the side, it somehow gets Carl Andre up from the floor again, just as Serra had lifted him up at the end of the Sixties.
Coincidental with the more recent changes was that Shapiro's working drawings began to appear, wonderfully labored drawings — black, smudged, corrected, and recorrected, the kneaded eraser and the razor blade and the pencil worrying the very fiber of the paper (Plate 25) — getting down, just right, those hard, elusive abstract-constructivist figures, right down to the very dust of the marking tool — the way Serra had pressed into existence the black paintstick drawings, making sculpture of them somehow, or as Bochner had worked out his gorgeous black and white arrangements of Pythagorean truths — the connections between these works are self-evident.
To make these connections pillories no one, least of all Shapiro, for

the connections of his work to Bochner's drawings may in turn be taken back to an acknowledged model in Jasper Johns — that's a strategy worth pondering — Johns, to Bochner, to Shapiro — though Bochner might be vexed by the rightness of Shapiro's discoveries, so free are they of any epistemic basis except for their chance "cruciformality," as Masheck called it. Bochner has borne down upon himself; Shapiro, by contrast, is just too unsure to let himself imagine that he really might be allowed to secure so privileged a place in the terms of the equation as I've stated it. But I'm talking about his head, not his art.

"You know, Robert," Shapiro said, "I've been toying with the idea of monumental sculpture." Of course he has — except for a vacillating disbelief in his own strengths, Shapiro has been a closet monumentalist all along.

*2 February 1978*

By far the great achievement of the past decade was the codification of the epistemic system, the ranging application of the system to the condition of style. The episteme — A-Z, say, or 0-9 or the spectrum sequence, RYB, or the days of the week or the months of the year — all those conceptual figments of truth and dozens more were at last seen as peer to the Realists' representation of nature or the Abstractionists' congruent presentation of form and content (Plate 27). To have imposed the epistemological vision as a stylistic equivalent to the other achievements was a great act of will — one in which LeWitt was accompanied by many other artists, though his retrospective must be viewed as emblematic of the overviews the others await. (And now there is this new odd sensation as well — as the battle has been won, so is the role of the supporting critic strangely egregious, an undesirable witness to the event, *la fee* Carabosse at Aurora's wedding, an embarrassment. At the beginning there is the artist, then the artist and the critic, then once more just the artist. But, in time, art history vindicates the critic too.)

For all its joy, the occasion was muted — since the imposition of an epistemic system as an authentic style in the history of art seems so over and done to me — certainly not for its initiators, but surely for those who came later, those who adopted it precisely because it was fast becoming a style among other styles, one that one might choose, just like that, out of a sales catalogue, as it were. But now that such a choice may be made, not from moral necessity but from taste, that there should now even be the possibility of a "Tenth Street Touch" with regard to epistemological abstraction, staggers me, hurts a bit.

Last night, we toasted the fact that the Museum of Modern Art had awarded credentials, *Lettres patentes,* naturalization papers, what have you, had bestowed on epistemological abstraction the irrevocable prestige that only the Museum of Modern Art can.

With regard to LeWitt specifically — I have always felt his use of the episteme was essentially intimate, best experienced in a scale dictated, say, by that of a textbook format, preferring the small drawings that bear the imprint of his fine draftsman's hand to the kind of work that partakes of a stagy declarative mode — those large reifications, the public metal sculptures, or the imposing wall works. So far as LeWitt is concerned, it is still the hand that carries the day, the little margins of transcending edge that the artist's drawing still mythically preserves — though LeWitt's epistemology expressly argues against my sentimental critical recidivism.

My reservations are linked to a perception that LeWitt is in no essential or fundamental way either painter or sculptor - a cryptomorph, a gynander, as it were, central for all that to the very aesthetic conventions of epistemic abstraction; instead, he is a draftsman whose insights are those of a born designer. That we should have come to accede that drawing is the halfway house between painting and sculpture and their peer, and that drawing is the essential embodiment of epistemic abstraction are the very functions of LeWitt's achievement. Yet when I meet a LeWitt in the guise of an extruded aluminum spray white "jungle jim," so to speak, or as a big Gulley Jimson of an epistemic wall, I feel, well, just indifferent to it. And I know my indifference is a form of risible atavism about what art is, or used to be.

LeWitt has amply and handsomely met head on all of the trying exigencies imposed by the new limits of his vision as to what art might now be. Acknowledging this somewhat tautological outburst, ought I not admire all other kinds of artists who unstintingly answer the imperatives germane to the arts as they conceive of them? I ought to, but don't. Certain ambitions seem so historically petty —so slim, despite the often fearfully difficult, virtuoso-like demands imposed upon the executant — Realism and sensibility abstraction, say. There are certain Realists and abstractionists at work today who generously answer the trying imperatives of their art - yet the boundaries of their Realism or their abstraction - not their art but the world to which their art is annexed — strike me as so petty, so bankrupt, in a word, so known — that for all my recidivist critical admiration for the hand, this admiration seems as nothing compared to the adumbration of a whole new system of visual order — and more, the imposition of the system as a style coeval with the earlier ones. And this triumph we owe largely to LeWitt.

## CAMP MEETIN':
## THE FURNITURE PIECES OF SCOTT BURTON

"In Burton one witnesses the force of convictions embodied within a finessing ironic parody." Lapidary, but too soon abandoned, this ending to a first essay on the artist. ("Scott Burton: Performance as Sculpture," *Arts Magazine,* September 1976, reprinted in *Post-Minimalism,* 1977.) Something more should have been said about the furniture pieces, but the works were not finished, or too few of them — just the *Bronze Chair* and two tables shown at the Whitney Biennial of 1975. But now that, in the last season, nine new furniture pieces were shown — four at the Droll/Kolbert Gallery, four at the Brooks Jackson Iolas Gallery, and one at the Guggenheim Museum (and each of these exhibitions filled with other already fabricated works) — there is now a body of work on which to build a surer assessment as well as to spur a broad range of speculation.

Only it is thorny. So many skittish issues are brought up by Burton's work, hard to straighten issues — especially since Burton's work, on its most obvious level, aspires to wedge between issues, all the while seeming very frank, very straight-out, very Square John.

From the outset, there is a certain casuistry in speaking of Burton's sculpture as either "furniture" or as "pieces" (Plate 28). Their generic grouping, as a broad consideration, suggests something of the theatricality by which Burton is drawn. Keep in mind that, in quite another light, he is an important force for a new style of theater and performance. An osmotic interface both separates and connects Burton's furniture pieces from and to his theater pieces.

Burton's furniture pieces shuttle between a double-edged postulate; while viewed and experienced as furniture and presented as functions of conventional ideas about furniture, Burton's furniture pieces, for all that, leap the very boundaries of those preconceptions, transforming them at the same instant into a meditation on furniture, on theatricality, on art, and on style. His work is both object and commentary at one go: sheer schizzy aesthetic autistic — he would have it — altruism.

Some descriptions of certain pieces: a Moderne-style table à la Milanese — blue granite exquisitely crafted, trifoliate corners, resting upon sledlike runners; or, a Constructivist/Suprematist table of Neo-Deco bent — square for the top, triangle as mid-support, and circle for the base. [The triangle edge extends one inch beyond the square's width — a dimension equal to the depth that the point of the triangle punctures the circular base — an intrusion equal to the thickness of the base itself. The measurements and surfaces and styles that are guyed (if, in fact, they are "guyed") reveal Burton's exquisite sense of proportion, elegance, and decorum — qualities that would more likely unnerve a Malevich through sheer hyperaesthesia than please him — quite despite the fact that the table is composed of the shapes in which Malevich discerned the presence of "pure feelings," the Suprematist triad, circle, square, triangle.]

At the other end of the style spectrum is a cast bronze outdoor table in the rustic Adirondack "camp meeting' " mode — a table associated, say, with the designs of Davis and Jackson, Downing and Walter — though, perhaps, it is met most frequently in its carpenter vernacular (Plate 29). Burton's version is single cast, unique; the wood prototype was lost in the casting. These objects — and the Guggenheim's *Children's Furniture Set* — continue to mark Burton's central place in the vanguard while demonstrating anew the degree to which the vanguard is concerned with and informed by past art — quite in opposition to the way conservatives behave, even conservatives of "modernist" stamp. Since the latter group "knows" just what art is, they haven't a care for vanguardism. How alone Burton seems then when compared to the prevailing abstraction, though his work is much influenced by its taciturn reductivism — one that is faltering, at any rate, even the best of it. Style gets used up — and, in time, Burton will also have to deal with that depletion.

All the time that Burton's furniture pieces are seen as furniture they are presented in a way that forces the contextual interpenetration of "art" — they are shown in a gallery, a special kind of signifier of sacramental or aestheticized space. Of course, the converse is also true — the furniture has the effect of reinforcing the initial profane identity of a gallery — that of a shop or store. Sure, that's difficult — since Burton's pieces retain their functional premise, a usefulness not eschewed even though shown within an art gallery or museum. His work refuses to take sides. It strikes a bargain between an accepted notion — the uselessness of art — while it argues against so received and unexamined a prejudice.

Beyond this, Burton's furniture exists as an amalgam of abstract considerations — surface, form, material, color — pure abstract information. This, plus a sense of learned historical style and paraphrase. [Before the square-banded legs of his black lacquer table, Burton suddenly grasped that what was coming through a lifetime of regional denial — a protest against feeling in any way Southern despite his Alabama Presbyterian birth — were the Greek Revival columns of the ante-bellum manse, church, and court house. Incredibly learned.]

Burton repeatedly worries at the hollow core at the center of art, teases the question of how art is a function of context as well as being its own formal self-realization (if it is really these things); how mannerism ricochets off a high normative mainstream, no matter how disdainful and unsanctioning that same tendency might be with regard to the reaches of irony that it feeds — and of which Burton is so arch an exemplar.

Still, Burton's art is not in the object no matter how fetishistically, how perfectly fashioned his objects. If the art is in the argument, one wonders, why even bother to bring its embodiment to so keen a pitch of finish — unless the work was all along meant to be a perfect object whose projection of beauty is a function of an indefatigable hand — Brancusi's ceaselessly smoothing way. In short, no one knows where the art of art is.

For all these "beyonds," these "yes, but" contradictions, Burton's furniture is meant to be used. There is the key. Burton has thought the matter through so much farther than the simple antagonisms I have pointed to; his speculations postulate a vacuum-creating third phase and, further than that, a democratic/populist "fourth dimension." It is these latter aspects and considerations, so marked by social mission, that transform for Burton the mere making process into a far more gratifying art-making one.

At the third level, simple erasure occurs — one occasioned by the incessant shuttling and blurring of context and content, in the playing of one side off against the other. On a certain level these contradictions are not even interesting any longer because they are so self-evident — to the artist at least. For the most part, the public may have difficulty even grasping this much. Burton thinks that the process of self-erasure postulated by a theoretical effacement of object by argument and back again conjures the far more important fourth phase — one that emerges to fill the void created by the preceding powerful frictions and, for him, obvious oscillations.

All this invokes a perhaps unexpected Whitmanesque streak in the artist, the idealistic belief in the democratic mind at its best, least sectarian application. For him, the furniture pieces, by a range of confluent processes still too fine to clearly analyze, aspire to some larger social and communal democratic content. Thus, the furniture pieces may be said to have their roots in his *Iowa Furniture Landscape* (1970). Here Midwestern storebought furniture was photographed in woodsy Iowa landscape sites.

Prompted by that seminal work — Burton says, "it just came out of nowhere" — I noted that:

> If one could trace the role of furniture in Burton's thinking it might run something like this: the objects of the landscape of the Iowa Piece yield up the notion of furniture as psyche; this in turn leads to furniture as surrogate people, which reverses itself into people-equal-furniture, a prefiguration of the *Behavioral Tableaux*. Last, if this system is correct, it recreates an entire system of psychic integration. Burton will be "made well" (assuming for argument's sake that he is "unwell") at the moment when the last equivalence is reached, when people equal people. I suppose I am describing a kind of therapeutic history, not a consecutive series, but one in which events happen all at once.
>
> — *PostMinimalism,* pp. 177-78

The elements of those analogies have been roiling now for over a decade in Burton's work: the object in the landscape, the furniture as psyche, the furniture as surrogate people, the people as furniture, the people as people. Such a graduated list informs the argument that inheres — for the artist at least — to the furniture pieces: furniture as furniture (use, style), furniture as art (the gallery context); the sheer obviousness and self-cancelling nature of a dialectical ploy in which neither primary term assumes control but which instead occasions phase three, a sense of "presence," that is, of objects of "persuasion."

This brings Burton to the most romantic belief of all: the vacuum that may be said to have been generated by an awareness of the intermeshed workings up to the third phase imbues his furniture pieces with qualities and values that as yet are but only hinted at, a sporadic description in the vast shifts in the sociology of art going on about him and to which he adds his special inflection, phase four. Through all of this Burton rejects all rhetoric or polemics. Such stances are, for him, a debasement, born of a confusion of rhetoric with "tone." Tone equals "presence" for him. "It is not what you start from," he says, "but what conclusions are drawn that marks the important critical act — for both the artist and the critic." The "fourth phase" — as if things could ever be so easily structured —equals a personal vitalization (how trite these phrases seem), the destruction, that is, of the fine arts notion of art-objects in isolation. In short, Burton aspires to the relinking of the so-called "decorative" with the "fine arts," or as I noted, ". . . the reintegration or re-assimilation of the decorative to art." *(PostMinimalism,* p. 179.) Burton believes that, perhaps, through the making of tables and chairs, at least the beginning of what he views as a "true populism" may occur.

The task is two-fold: for the artist the formal task will ultimately work for the creation of objects in isolation to the formal task of the entire space — of wall, floor, environment, the whole decorative ensemble; for, as Burton rightly sees, "the whole space equals the real unit of social history."

But, on the other hand, perhaps an even weightier question concerns the nature of the spaces being or to be provided; what spaces are given to be serviced — are they ceremonial or sacramental, domestic or commercial, virgin nature or reclaimed territory? Perhaps here is the most critical consideration of all - though first steps have been made in the tables and chairs: what he is really dealing with has yet to be given to get done.

Burton is no woolly democratic zealot. He is far too canny and hard-headed to succumb to the contemporary delusion that artists legislate change — viz. Joseph Beuys, Joseph Kosuth. Still, the wild thing about the furniture pieces is that they —certainly for the artist — approach his fourth phase of consciousness, one which taps a broad range of artistic and social connections that are as similar in spirit to, as they are dedicatedly different in appearance from, say, an earthwork by Robert Smithson — to take the example of a sculptor totally animated by an Olmstead-like and Whitman-like vision of what American art ought to be. What Burton's furniture pieces abolish, obviate, or, at the very least, momentarily staunch, is the powerful and determining grip of shabby etiquettes, labels like "mannerism" or "realism." I suppose what Burton's furniture pieces make me see — having come this far with him — is that the issues he hopes to address are issues beyond style, styles, or even stylishness.

# BENNI EFRAT: THE WHITE RECTANGLE

*2 May 1976*

If one isolated the paradigmatic image haunting painting from the mid '60s through the present, it would be the white rectangle — both adversary and beloved — the preordained object of the creative impulse (can it be newly imaged?) or the ultimate crazing image which, even as it is destroyed, or in its destruction, expresses a tenacious supremacy.

*Painting* (1975), a film by Benni Efrat, shows an image of the artist in the act of painting a white rectangle in firm, sure strokes — first across to the right, then down, then across to the left and, last, up — creating a brushy white rectangular field. The screen upon which the film is being projected is the very same white rectangle the painting of which is the subject of the film. In *Painting,* the perimeter of the white rectangular field is congruent with the white rectangle of light, thrown directly upon the white painting which had formed the original subject of the film.

Another work, *Round-a-Round* (1972), suggests Dibbets and Snow, though its early date throws neat stylist interconnections and sequences into doubt. To describe it is a bit taxing. The film is projected in side-by-side rectangles. The rectangle on the left is a black and white film and shows the artist walking in a circle, turning a large wheel upon which a camera has been mounted. This large wheel is affixed by an axle — a constant radius — to a central pole. Hence, two circular paths of movement are created, the wheel itself turning in a circle and following a circular pattern. These two orbits are recorded in color by the camera mounted on the wheel. The right-hand screen shows the color film taken by the camera recording these movements.

Still another work of 1973, *Tailgate,* records the movement of a car being driven through London tailgated by a second car. Car 1 throws out a brilliant light from its rear window. Car 2 picks up this beam with a camera mounted in front. That means that when the second car is directly behind the first, the image is brightest — a beam of light directly into the eye, so to speak, of the beholder. Thus, variations in gray (if you like, still another definition for film) occur when car 1 turns to the left or right as it weaves through urban traffic and car 2 tries to follow immediately behind.

*18 January 1977*

A darkened room set up with pedagogic blackboards that serve as screen, paraphrasing the whole range of the expressive didacticism of the institutional learning process (student equals spectator, picture equals blackboard, artist equals instructor): Benni Efrat's new 1976 films shown at the Whitney.

In a way, were I to string together journal entries going back to *Les Cousins, L'Avventura, The Chelsea Girls,* and Efrat's *Putney Bridge* (emblematic of the five films Efrat showed), I would have adumbrated, as it were, an article encapsulating, I believe, a critical suite of filmmaking from the late '50s to the present, expressed naturally enough through my biases. These were among the films that most stirred me as I entered my years of critical susceptibility.

Lurking behind Efrat's films is the central premise of the white rectangle, invoking a most elaborate irony about the beauty in, and the futility of, ever hoping to surpass the primordial moment of modernist art history, the discovery of the First Suprematist element. Yes, yes, that was a black square upon a white ground (that is, until 1918 — when Malevich painted the great *White on White*). And even here, that memory is alluded to as Efrat's screens are *black*boards.

The first film of the evening, *Pastel* (1976), continues the pure information of the kind that marked Efrat's earlier films, those he showed me back in May. Here a color film image is projected of the artist marking parallel columns of white chalk from left to right on a blackboard. Upon the projected film image the artist literally re-enacts this procedure so that as the literally real new chalk columns appear on the blackboard, the film image becomes sharper, more luminous because it is being projected upon a new white ground, until, of course, the whole blackboard is covered with white chalk.

As the film ends and the white light trailer is projected on to the board, we now see the brilliant image of the now real parallel chalk columns.

It no longer is clear when the conceptual performance begins for, prior to the projection of *Putney Bridge* (Plate 30) upon this blackboard, the artist takes a sponge and wipes away the chalk columns. And suddenly, upon the wet chalk-smeared blackboard, the image of *Putney Bridge* is projected — the bridge, the Thames below, moored boats, occasional figures along the diagonal right-hand *repoussoir;* all this takes into account Monet of the early and late London views from the 1872 *Impression: Sunrise* — from which the Impressionist movement receives its name — to Monet's post-1900 Thames views with their consistent high horizontal bridge motif. This allusion to Monet and to the *Impression: Sunrise* is rendered virtually explicit when, as a function of the dramatic narrative detail, a scull filled with oarsmen goes by, interrupting the somewhat more low-key activities of the figures relocating themselves along the diagonal embankment. That these latter figures' main occupation is skipping stones suffuses *Putney Bridge* with an unexpectedly nostalgic cast. Across these river images, the artist chalks and smudges patches on the blackboard — small areas of chalk against which the pale projection becomes, by contrast, stronger.

Through a purely fortuitous moment all modernist art history was gathered up. On the blackboard, when *Pastel* had been wiped away, minute rivulets of water had coalesced so that, at last overcoming the inert holding power of the chalky powder and gaining sufficient energy from simple gravitational tug, they formed a sudden dark drip — the insignia of Abstract Expressionism — across the face of the blackboard, a calligraphic rain trace. Here it was — the birth of Impressionism to its apogee in Pollock — all conducted within the framework of a conceptual epistemic film performance.

Were this not enough, the artist then proceeded to another blackboard and *Cement Plant* began. Again, the film referred to the emotion-laden simplicity of an essentially horizontal landscape upon which the artist slowly picked out errant details; smudging, scribbling upon the projected landscape image, every now and then indicating a row of smokestacks accruing in the landscape like an industrialized pollution. I thought of the industrial incursions in the landscape imagery of Neo-Impressionist landscape painting — the horizon factory towers in Seurat's great *Une Baignade à Asnières,* for example.

The fourth film, *Neta,* is named for Efrat's friend, a beautiful Israeli dancer now working in Switzerland. Efrat filmed her in the act of a very careful application of makeup. As she applied the various elaborate stages, Efrat mimed the gestures with chalk and pastel sticks upon the projected image. Foundation color, face powder, eyeliner, eye shadow, false eyelashes, liquid blush, lip rouge, etc., etc. However, these actions of applying makeup, while illusionistically "real" in the film, were re-translated back to a literal two-dimensionality as they were re-enacted by the artist's hand upon the blackboard. As the film ends and the white light illuminates the blackboard, one sees the memory of the 1870 Degas pastels of ballet dancers transformed, as it were, to the 1950s studies for *Woman* by de Kooning.

The last film, *Matter on the Move,* invokes once again the more non-associative epistemology of the earlier films. Here the artist white-chalks in upon the blackboard a growing checkerboard of white, red, yellow, and blue squares with a somewhat Dibbetsy-upsmanship, for as the columns accrue, the artist in this film tips, in varying degrees of acuteness, the original colored columns, while the real artist present in the gallery continues the increments of white chalking-in over the projected color patches. This process continues over five columns tipping left and right.

In these films, more properly, in these film performances, bridging the projection of past action and actual re-enactment of that action, the tension is so acute that one imagines still another film — that of the artist enacting the alteration process itself — I am talking about a film which would be a film of a film — ultimately, a film of a film of a film ad infinitum. This endless column of image, while perhaps possible to film, is essential, germane, and organic to video, so that Efrat's films about painting are, in that sense, a paradoxical interface between film and video.

A last detail (I suppose privileged information). The catalogue issued by the Whitney in connection with these films is dedicated "to the person I never met" (and that of the concurrent retrospective of paintings and sculptures at the Bertha Urdang Gallery: "to the artist I never met") — a sentimental crunch that repels me. Risking rebuff, I question the artist — Efrat recalls an exhibition a decade back held at the Louvre, a show of *peintres inconnus de la Renaissance.* One work by an unknown master brought him up short — a large figure composition atypically non-central in organization — a crowd in the lower left-hand corner played against an empty upper right-hand space. One of the figures looks directly out of the picture engaging the spectator's glance. Taking into account the iconography of portraiture, Efrat rightly assumed the figure to be a self-portrait of the artist himself. In a powerful surge of identification with that artist, Efrat felt indicted in his own dubious painting — he was still "a painter" at that time — and, Saul into Paul, he resolved, at that instant, no longer to paint. Ironically, this resolve bridges the five centuries between a Renaissance self-portrait and the modernist obsession with the white rectangle.

# SIAH ARMAJANI: POPULIST MECHANICS

Siah Armajani is a wiry, gracious man punching forty, small of frame but not slight — dark hair shot with gray, a nose flattened and askew, but, for all that, a handsome Persian. His conversation is animated and generous. He wants to tell you everything and not leave anything out. For all its learned points of reference his conversation is never pedantic or idly name-dropping; it never calls up inappropriate sanctions. Each of Armajani's references — and there are many — have been assimilated in an experienced, authentic way.

His people are Moslem Iranians with admixtures of Christian and Jew, an important detail suggesting the tolerant atmosphere in which he was raised. His father and uncle were both mission-educated — discreetly. It was (and, in even greater measure today, still is) of dubious form to be western-educated. Current Arabist chauvinism is especially marked by the activist, expansionist nature of Sunnite Islam, so different from that of the less-wordly, bookish Shiite sect from which Armajani stems.

Armajani's uncle Yahaya emigrated to the United States becoming a professor of history at Macalester College in Saint Paul. Through the avuncular good offices, Armajani was in turn able to come to this country. He recognizes that his humane populism - he believes in the natural and evolving goodness of humankind in a way I find difficult to accept — as an artist and thinker marks him as alien to present-day political repression in Iran and he entertains no thought of returning there. In 1967 he became a citizen of the United States.

Partly the result of his work, partly that of his sheer disarming force of manner — Americans are more apt to countenance Jeffersonian ethics when uttered by outlanders — Armajani has been party to a wide change throughout American sculpture where its most progressive expression leans toward an unapologetic construction, not an aestheticized Constructivism. Armajani lurks at the edge of this development, perhaps without receiving the full share of praise he deserves for the extraordinary effect he has had — although Linda Shearer's inclusion of the sculptor in her exhibition *Young American Artists 1978* (The Guggenheim Museum, May-June 1978) and the early supportive writing of Michael Klein, notably the essay for *Scale and Environment: 10 Sculptors* (Walker Art Center, 1977), made inroads in dispelling Armajani's neglect on the East Coast. But to suffer from neglect — or, more accurately, not to yet be part of canonic taste — is equally the lot of the very sculptors with whom Armajani is reasonably associated — say, Michael Hall at Cranbrook or Edward Levine latterly of Wright State, now Dean at the Minneapolis College of Art and Design, where Armajani has taught for years. And behind these names, Armajani's included — though in varying degrees of absorption — stands the moral, if not always formal, example of Robert Irwin, as stirring for Armajani, quick to acknowledge his admiration, as was the impact made by the American Patrimonial texts Armanjani so ardently committed himself to on arriving here.

Doubtless one thinks that I overstress the political content of Armajani's work, especially since — on a certain level — one really can't see it. But any claim for form — from the most empirical to the most transcendent — is invisible. Still Armajani's work is virtually ununderstandable without reference to his political awareness, one that leads him to make an abstract sculpture embodying, according to the artist's view of the matter, a special take on the long suite of Enlightenment philosophy from the Rationalist ethics of the eighteenth century to the structuralism and epistemology of the present day. On its simplest level, form for Armajani is iconographically predetermined and embodies a populist democratic ideology and frame of reference. Yet his forms are never explicitly illustrative; or even if in some sense they are "illustrative," they are never "narrative," this despite the enormous number of philosophical and literary references his sculptures make, at least in terms of their titles *Thomas Jefferson's House . . .*, *El Lissitzky's Neighborhood . . .*, *The Red School House for Thomas Paine* (red school house, a nostalgic symbol for a lost pre-industrial America; red for the "radical" views of the revolutionary pamphleteer), and so on.

What is most immediately visible in Armajani's work is its absence of conciliatory sweetness or tactile appeal — it ain't smooth — physically or conceptually. This indifference to — even rejection of — the merely pleasing is realized through disruptive passages, bumpy elements, frank abutments, blocked approaches, "wrong" heights, makeshift executions, obvious colors — all sorts of diffident discrepancies that force a mythically desired aesthetic "whole" to fracture into a valuable particulateness; and his work is large, always as big as a house or a barn or a bridge — or even larger. Klein describes, for example, *A Fairly Tall Tower: 48,000 Miles High and The Ghost Tower:* "The former is envisaged as a self-supporting tower suspended in space, anchored to, but not supported by the earth. The latter was a holographic image floating above the First National Bank of Minneapolis. An inverted mirror image of the bank, the hologram would appear and disappear, depending upon cloud conditions and would seem to extend 382 feet into the air." Klein goes on to describe still a third project of 1969, *The Tower: North Dakota:* "For this project Armajani calculated — again with the aid of a computer — the dimensions and shape of a giant tower that would cast a shadow across the state of North Dakota" (*Scale and Environment,* pp. 16-17). That Armajani recognizes that these conceptual projects paraphrase, say, the Eiffel Tower, Tatlin's *Monument to the Third Internationale,* and Frank Lloyd Wright's *Mile High Skyscraper,* goes without saying. But in what way such paraphrase is stylistically linked to these models is unclear. Formally? Sure — and yet there is that oblique, literary nerve center, that punning verbal/formal side of things in Armajani. One can't help, say, chuckling over Armajani's *Boat/Bridge* (1969): two feet short of the width of the river it must cross, the traveler enters one end of the *Boat/Bridge.* Having sailed but two feet, the traveler disembarks on the opposite shore.

To be in an Armajani (or to look at an Armajani) is to be inside a big structure that isolates and atomizes rather than harmonizes or aestheticizes. Armajani's part-to-part legibility marks a difference from the smooth sailing, the even flow of part to part, or part to whole, provided by some of the very sculptors upon whom he has exercised, at least in part or in passing, a certain appeal, let alone of the entire Constructivist tradition. And for all the scale of his work, there is nothing engineered in an Armajani, despite an often engineered look — nothing that smacks of high technology. Though Armajani may have had recourse to complex machinery — indeed had — it all looks as if it had been made with simple technology and tools, not complex ones.

For Armajani, sculpture is big, built and experienced in separate episodes — and, unlike David Smith and his suite, it is never cryptically anthropomorphic. Each of Armajani's forms is unrepentantly clichéd — clichéd colors (the red, black, white of Suprematist graphics, a favorite reference); clichéd building material (schlocky two-by-fours, corrugated tin, shabby transparent plastic, all chosen for populist associations); clichéd space (room, corridor, window, pen). Ironically, his sculptures revoke the very American notion that form follows function — taken for a Bauhaus adage but one initially drawn by Horatio Greenough in his *Observations and Experience of a Yankee Stonecutter, 1852:* "When I define Beauty as the promise of Function; Action as the presence of Function; Character as the record of Function, I arbitrarily divide that which is essentially one." By contrast, in Armajani's work, forms connect to specific pieces of information, specific spaces, specific materials — but as to function, there simply is none. Room-like spaces are not passageways meant to lead you somewhere else. Forms and functions accrue not from spatial necessity or any kind of felt urgency (the doctrine of early abstraction), but because there exists a kind of guide, some builder's manual, some popular-mechanical point of reference of which this space is merely a sample. His is a kind of flatfooted, Benjamin Franklin, Show-Me-I'm-From-Missouri, literalist empiricism — literal matter, literal/verbal. Indeed, Armajani's most intensive literal/verbal work may have been the creation of a kind of builder's reference guide in which all possible combinations of space are indexed — a "room," say, with a "window" on the "wall," in the "ceiling," or on the "floor" — similarly one with "doors," flat on the floor, standing up the wall, open, closed, inside-out, top and bottom. In such a sense, aesthetics doesn't even come into consideration. It is a kind of architectural sculpture by rote. He calls this encyclopedic manual, and he has made it, his *Dictionary for Building* (c.1973-1978).

Armajani ardently rejects any type of Palladian centricity, even in the Thomas Jefferson houses where it would be natural, as too patent an aristocratic class signifier, one often superficially elaborated in middle-class architectural spinoffs — chintzy, plastic symbols of upward mobility. Armajani understands this as well as Robert Venturi, without projecting the latter's enthusiastic, slumming condescension. He rejects time-honored aesthetic considerations — more interesting perhaps in the degree that they are ironically imitated — for the blunt proposition that structure merely encloses or discloses, but not necessarily in any hierarchical or aesthetic way locateable to taste decisions — though, in the end, this too becomes another taste.

Armajani postulates a sculptural transition from inside to outside, resulting from an ironic commentary on philosophy, one realized in the simple forms and actions of a carpenter vernacular. The reason that this vernacular is so important to the artist is that it itself has come to signify pragmatic, democratic self-reliance. Additionally it provides a way of tying the American expansionist building types of our Manifest Destiny — carpenter Gothic, shingle style, pseudo-Queen Anne beach cottage, boom-town balloon frame and Main Street facade, etc., to a populist-derived architectural theory. This theory, really a "post-modernism" as architectural historians are coming to say (Robert Venturi, Vincent Scully, et al.), is coterminate with the abstract formalism of the Suprematist/Constructivist ethos. There you have it. The balloon frame of the Midwestern town transformed into Tatlin's *Monument to the Third Internationale.* Only in America. For Armajani, the question whether such principles can lead to beautiful sculpture simply (Plate 31) never arises — and in not having arisen, leads to beautiful sculpture.

Armajani's focus on experience, sifted from the texts of the Founding Fathers through the Transcendentalists, thrusts him headlong into a form of pure Phenomenology from Husserl into Merleau-Ponty. The isolation of the physical fact is the real intent of his sculpture, and to explain this Armajani tells the parable of the hammer taken from Heidegger's *Being and Time:* in hammering away at a nail there is but one consciousness, one psycho-physical experience — one's focus on the hammer driving the nail through the board. In this way, experience becomes whole and continuous; as such it is in touch not with the present but with the future. Thus, through wholeness and continuity, the experience is transformed into an aesthetic and kinesthetic episode. But what happens when the hammer breaks, asks Heidegger. Armajani happens. Instead of continuousness of movement, one is faced with an experience composed of isolated parts devoid of a sense of futurity: to wit, the hammer, the nail, the board — alone, in and of themselves. The continuous drive to a future — seamless and unawares — that renders such a building experience aesthetic is rescinded in Armajani's way of building. To regain the sense of each instant, each part, but *not* to bring one to an aestheticized future, but rather to the consciousness of the intensity of an ongoing present — such is Armajani's ambition. And achievement. Only the present is real for him. So he is drawn to the quotidian banal, the simple, prosaic, and the over-familiar; through the ad hoc working of grungy physical matter, he keeps one honed against a keen present. The truest dichotomy of Armajani's work is not that of the disassociation of art from nature, but between doing/being and the sense of aesthetic futurity. In achieving this he replaces a shopworn sense of sculpture with a fresh and, dare one say it, American one.

# LUCIO POZZI:
# THE CONTINENCE OF LUCIO

*13 November 1977*

Pozzi =wells; Pazzi =Crazies. The Window Series at P.S. 1 =Lucio (light), Pozzi (wells) =Wells of Light (Lucius Wells?); or are these wells of light a crazy version, a secular Pazzi Chapel (Plate 32 and 33)?

Born to a family of the high bourgeoisie whose now-reduced fortune originally was based in North Italian textile production. Four generations of Ajroldi-Pozzi forebears: his great-grandmother wove beautifully and set up a single loom. She took on a girl and a second loom; at length, six girls, six looms, and then a dozen of each. During the grand paternal generation — industrialization. In a short decade, Pozzi's grandfather on his mother's side rose from sales representative to administrator (among other things) of the Teatro alla Scala in Milan, dying before the war and the destruction of the opera house during the Allied bombing of Milan.

An isolated, awkward teenager, the artist first saw himself as a poet in the manner of a bilingual Verlaine; there was a lot of il-pleut-dans-mon-coeur-stuff, and the ardent, if blaspheming Catholicism typical of the style — lyric, ejaculatory appeals to Gesù Gesù.

Pozzi's mother Ida, deeply artistic, perhaps was embarrassed by her gifts and, as a good bourgeoise, while dissipating energy, always made substantial contributions along the way. She founded, for example, a circulating library, Il Carosello, in Milan following the war, bringing home a wide circle of literati who encouraged the youth to continue on as a poet.

Lucio Pozzi's studies were disastrous. For all his denials, though, the humanist traditions of North Italy are still the bases of the artist's wide-ranging cultural referents. The dormitory of his boarding school is still clear in his mind and remains a kind of Pozzi *in petto* — a room of a nineteenth-century villa in the Renaissance style, a bare room; outside the window stands a mimosa tree, brilliant yellow in its springtime bloom — and beyond that, the blue Gulf of Genoa. All simple, direct, urgent. Like a Pozzi. On one wall of this room are the art postcards published by the Galleria il Milione, after the work of early twentieth-century European masters — as you might expect, but Arthur Dove and Marsden Hartley, too — abstract, ideogrammatic, compelling scrupulous examination from the young boy.

Pozzi's first pictures were made during a childhood quarantine when downed with scarlet fever. His nurse, by way of diversion, showed him how to paste collages together and to paint in watercolor, the latter a medium he has never abandoned.

Then, too, there was the extraordinary sideswipe of the Scottish sculptor, Michael Noble, the artists's mother's second husband. Noble opened the family's villa in Lago di Garda to friends from his Paris days, friends from the Diaghilev circle and the ballet generally — Russian emigrés such as Igor Markevitch or Boris Kochno. Pozzi remembers his first abstract gouache attempted an ideogram of a ballet dancer, abstracting along the simplifying lines to be seen in Noble's sculpture. This ideogrammatic side of things also has a curious connection to Balla, a not gratuitous linkage as will be seen.

Noble's associations with Italian art, too, brought Pozzi into intimacy with the traditions of Italian Modernism. Especially close to the household was Noble's dealer Ghiringhelli, then the director of the Galleria il Milione. On one hand, the latter defended the traditions of classicism associated with the Scuola Metafisica — at this date Morandi was its great exemplar, having supplanted the vulgar late de Chiricio (Plate 34). Ghiringhelli championed the experimentalism of Futurists of all stamp, Sironi say, or Magnelli, even late Severini — the latter phase of Futurism so contrarily and perplexingly absorbed into the classicistic character of Italian art of 1920s through the '40s, into the work of the *Novecentisti,* that is the "Painters of the 1900s" (as Sironi and his colleagues were called by the Fascists); or even more importantly, the great sculptor Arturo Martini, whose work afforded Noble a model for his own fusion of Moore and Giacometti; a not antithetical mix, when you think about it. Then, too, Ghiringhelli was aware of aspects of the *informel* developing in Paris following World War II, leaning more naturally to De Staël rather than to the Fautrier side of the question but *informel* for all that.

Ghiringhelli always found an hour to visit the young painter, telling him how he remembered Kandinsky would move a brush — "like so," doing it Kandinsky's way (since he had been a friend as well as a painter in his own right); or how Morandi would put down a color or make a line — "like so." All this was unself-consciously performed, just part of the distractions of the household, slightly amateurish even, if you see what I mean.

Still, in all of this Chekhov-like indolence ran a serious strand. Noble and Pozzi's father, after whom the artist is named, believed in the supremacy of drawing (for quite different reasons), of the importance of conquering anatomy and perspective, that kind of thing. Thus, for all the advancement of the experimental espoused in the villa, there would also be that solid understructure so typical of an early Modernist view of abstraction - its derivation, that is, from a design-oriented simplification of nature strictly observed. The traditional strain of a businessman-father naturally would lead him to believe that his son's development as an artist should also be made along conservative lines. So Pozzi, ardent in everything, lived with a human skeleton in his bedroom, drawing the bone structure and inventing the concomitant musculature.

As a young man then, Pozzi's cultural fusion was that of all high *liceo classico* culture, one marked by a lyric expressivity far deeper than that permitted by the social norms in which it was set. This "Leopardismo," if it may be called this, is shaped by the recalcitrant tradition of Pozzi's Lombard forebears; culture was humanist, understated, liberal, consciously resisting the ultra-montane traditions of Catholic Letters. It maintained a quasi-heretical stance, blonde in its darkness, Protestant, Arabizing (in the manner of Averroës) if not even a kind of Hebraicising Catholicism. (That Pozzi's son — he also has a daughter, Eve-Anna ["Nina"] — would one day be called Giordano after Giordano Bruno is what I mean — classical, mystical, heretical; it's all there — the spiritual forebear of Spinoza martyred by the Church for what we today recognize as an early expression, however mystical and heretic, of the modern tradition of philosophical inquiry.)

At twenty, Pozzi began to break from his family, enlisting in officer's training school, a choice based on class and costume — officer's white gloves, you see. His military service enlarged the range of his national awareness, lessening the emphasis on the Lombard and the Milanese, pushing him into the Roman arena. It was necessarily South Italian too, a kind of inescapable earthy testing that the experience of the military reinforces.

Since officers were allowed greater freedom, Pozzi's architectural studies began at this time as well. They were highly colored by the nascent *Partito Radicale,* a movement particularly felt in the Italian architectural schools of the day. The *Partito,* to all intents and purposes, directed the national student movement. It was in the architectural schools that the early rebellions — culminating in the uprising of 1968 — were first manifested. An early student rebel, Pozzi and a fellow architecture student once refused, for example, to execute a simple Beaux-Arts rendition of a column in a Greek order. Pozzi and his comrade insisted that the social symbolism of the exercise be disclosed prior to making the drawing.

During the years of his architectural studies, Pozzi was a founding organizer of a co-operative of graphic designers whose ambition was to improve society through the solution of graphic design problems — the models of Werkbund, Bauhaus, not to say Kelmscott and Jugendstil run before him in this systematic utopianism. Pozzi grasped his elitism a bit better when the text of his first poster (one set upside-down) was merely shorn off by the socialist cadres to which it had been sent. Pozzi had intended that, when inverted and alternated, the poster would form a new coherent mural-like sequence. The Socialists merely thought that the line of propaganda had been incorrectly set and printed.

In Rome, too, he discovered that he was persona-grata in the city's literary and artistic community — Moravia, Elsa Morante, the young Pasolini. Pozzi's first wife's family, distant Sicilian nobility, moved more or less in the same social sphere of his own, and the marriage, though it at length failed, was at first thought to be quite perfect.

His mother-in-law, Topazia Alliata, had opened the Galleria Trastevere, one of the period's most advanced galleries. Ambivalent about the Italian *informel,* she went so far as to scandalize her artists by showing, say, the young Piero Manzoni, just beginning an intensely iconoclast career, one that in its own way despised painters and painting, that is, despised all that Pozzi was drawn to and believed in — as he continues to believe — despite all his curious philosophical detachment from conventions surrounding painting as an end.

It was to the Trastevere that Lawrence Alloway, then the ambitious director of the Institute of Contemporary Art in London, brought the young English painters who worked around William Turnbull — Marc Vaux and Gillian Ayres among others — painters who had absorbed the influence of Rothko, Newman, and Kelly before Italian painters had availed themselves of these now preeminent models. Pozzi's lean and simple solutions led him directly to Kelly's work, because they were both working at a point of extreme simplicity, though in disparate ways; and when at last in the United States, Pozzi sought out Leon Polk Smith, still one of the least appreciated American abstractionists, though one always championed by Alloway.

Pozzi's world then was curiously schizzy. On one hand were the literati whom he frequented — naturally dominated by Moravia. The latter intensified the artist's social conscience, informing Pozzi, as best he could, of the Italian role in the Holocaust. The literati all read Wittgenstein and Adorno. Then there were the architects: always sloganeering, theoretically and philosophically inclined. Their catch-phrases of Heisenberg's indeterminacy and the mathematics of Gödel also provided for a new emphasis on aleatory structure. But architecture then mostly meant "urban design," while aleatory meant music.

Through all of this Pozzi intensified a penchant for all things American. Always a passionate enthusiast of jazz, he now discovered the "indeterminacy" of Cage, which he regarded as a type of "indeterminate seriality" (if there can be such a thing) paralleling the musical seriality of Karl-Heinz Stockhausen. He recalls an underpopulated Festival di Venezia when he was one of the rare audience members to remain to the end of David Tudor's and John Cage's concert of the latter's music. He realized then that his lot was to be cast with the Americans, slowly abandoning a community and a society that he came to recognize really had despised painting despite lip service. By 1962 he had tentatively set himself up in his first New York studio on 4th Street and Second Avenue.

The earliest work Pozzi felt secure enough to exhibit publicly was painted at the end of the Fifties. These works, large by contemporary Italian standards, hence "American," were broadly brushed simple arrangements. Often as not, these emblemlike configurations were so thinly painted as to reveal bristly striations of stroke and wash. In this sense, they are not without a certain oriental cast.

At the time, Pozzi was committed to a bright, emblematic figure, so large a figure in fact, that the even larger format of the canvas seemed to cut the compositions arbitrarily. The painting is marked by a sense of scale, a flat designer-like configuration, an impertinent color, and in an occasionally soft-modulated profile, an organic connection. Works of this type were shown in 1963 at the Galleria l'Indice in Milan. Oddly, had these works any creamy thickness they could bear comparison with contemporary paintings by Al Held, those swelling variations on geometric shapes that occupied Held until he forswore the use of color in 1967.

A curious subtheme of Pozzi's essentially Abstract-Expressionist position at the turn of the decade is to be found in simple constructions based on prefab-modular components, cast-iron pipes, say. Presciently, Pozzi sought to contrast "very much of something" — that is, the big painterly canvas — "against very little of something else" — the painted pipe sections. And in so doing he stumbled, as it were, into a kind of talismanic minimalism, the private reductivism of, say, certain very early objects of Judd, LeWitt, even Bochner, not to mention even more obvious affiliations to Tuttle. All of this will be clearer much after the fact; at this point, I believe these odd private objects serve a kind of neutralizing, even possibly curative role vis-à-vis the public, clamorous character of the artist's preeminently painterly concerns.

In 1964, the year Pozzi definitively settled in New York City, the abstract painting continued apace, scale and tart color anticipatably still in the forefront. At this point, the first biomorphic stirrings adumbrated in the earliest shown paintings become a fully-explored, even intentionally vulgar transposition of Surrealist biomorphic accretions (Plate 35), ranging in character, say, from the high art associations of Arp to the down-home biomorphic aggressiveness of Walt Disney (or earlier, the animistic cartooning of Charles Burchfield). This shift was facilitated by the period's most characteristic change of medium — the giving up of oil paint for acrylics, the physical properties of which allowed for the broad biomorphic resurgence in painting at the time.

By 1968, impeded by the despairing turn taken in the world at the time, Pozzi, like many, abandoned large-scale painting, giving over most of his energies to teaching and to the small, notational watercolors he had always kept as a kind of picture album of his life. In the last months of the year, the scale of Pozzi's work diminished radically. Rejecting acrylics as somehow basely materialistic and inculpated the by the failure of the American Imperial adventure of the day, Pozzi's last works were tiny gouaches whose shapes were somehow lost in gray fields. These feckless, vestigial elements were "left-over" after the "very little against very much." In a certain sense, as the imagery had diminished to near nothing, so too had Pozzi's idea of painting itself. But if diminished, it was still there, near dormant. The painter had not really fully accepted the prevailing studio cant of the day, one that precluded painting at all.

It can hardly come as a surprise to those who follow the sequence of contemporary art that 1968 represents a most novel and critical fulcrum for the shift in priorities of sculpture and painting - as important as, say, 1912 or 1947. It is then, and for the subsequent four years or so, that the classical figures of post-minimalism were producing their most recalcitrant and unyielding work. At this most critical juncture, Pozzi too felt that painting as it hitherto had been known was somehow ethically inadequate, an insufficient position. And like most, he too has by now returned to a less distressed anxiety about painting as such, so that what is presented in his paintings today seemingly may be assessed according to more conventionalized systems of appreciation. The widespread resurgence of painting that we associate with the present moment, though, is of a quite different order in the case of artists of an initially conceptual bent from those paintings which derive in an unbroken line from the easel conventions of modern formalism. The latter stand quite nakedly as merely a new academy; the former must be seen as an expression of a continuing if deeply modified organic relationship to its conceptual origins, rather than as merely more "painted things" in the world.

Among the provocations so indicative of the post-minimalist break of 1968-72 is the emergence of a behavioral and performance-oriented manifestation. By April of 1971, Pozzi found himself at the turbulent center of this startling development.

In connection with an exhibition of paintings held in Italy, Pozzi created an aesthetically polarized situation. Six close-valued taciturn works were countered by six strong declarative performance operations in two adjacent spaces at the Galleria dell'Ariete in Milan. "I don't mix incompatibilities — I just juxtapose them."

Whereas the New York vanguard had already been startled by a wide array of performance-oriented art (Vito Acconci above all, which in part continued the submerged tradition of the Happenings of the 1950s), Italian manifestations had lost touch with the protean originator of the Happening, the Futurist performance. Perhaps the need for Performance Art in Italy, that is, from a certain ironic viewpoint, unexpectedly had been answered by an intense political life played out in the street, workers' manifestations of all kinds, not to mention the linked kidnappings — political or brutishly extortionary — and in the end the assassinations. Nothing like this disturbed the admittedly jarring streets of New York.

As it happened, Pozzi's performance pieces in Milan coincided with the first judicial assassination there (Plate 36). Pozzi's personal, even quietistic, performance (suggestive perhaps of the role that George Brecht played in contemporary *Fluxus* performance in Germany and France) was announced as a public execution, a civil killing symbolized through the modest gesture of dropping a red carnation after an actor/victim was shot with a blank cartridge. Needless to say, the agitated state of the police led to Pozzi's arrest and jailing. After two days of friendly phone calls between family friends, Pozzi was released and he resumed the announced performance activities — to paint in the gallery "on order" and to sell the results for a dollar apiece; or, in homage to Hans Haacke, to request gallery visitors to fill out questionnaires about their attitudes toward Nature, Art, Politics, and Sex. All this activity and its considerable multiplication over the next few years was part of Pozzi's need "to do as many things as possible different from painting — to show painting as just another thing — yet always to refer to painting."

Truthfully, the juxtaposition of performance and straight academic abstraction was not especially popular at the time. It still isn't, since the work was neither sanctioned by painters nor the nascent conceptualists either in Europe or in New York City. Pozzi's dismissal by these various groupings, the condescending silence that marked the reception of his work, further intensified his commitment to a perhaps untenable position, merging also, perhaps, with a certain undertone of reactive criticism that marks his work to the present. That painting and performance-type activity did not seem to marry well in the studio context and not at all in the gallery — this confrontational mésalliance — perversely was agreeable to the artist in the degree that he felt that "Things just belong to each other without having to worry about whether they ever did."

But the irresolution nonetheless sparked a period of retrenchment, even a kind of continence during the very period of the most radical art activity in New York, an activity of which Pozzi's work is a very natural product and expression. In 1971 Pozzi, who had been teaching at the Cooper Union, organized a structured performance activity of 29 nude models. The work, played throughout the institution, was highly nervy and set up in a gamelike way with Pozzi calling the shots. If Duchamp's *Nude Descending A Staircase* is annuciatory of the death of painting, as it is said to be, then "*'29 models' Descending an Art School Staircase*" is an even greater mutliple of calamitous absurity (Plate 37). At length, Pozzi was asked to leave the faculty, though he now teaches in the perhaps more anarchic ambience of the School of Visual Arts. These encounters with the conceptual granted, Pozzi's main preoccupation remained dedicated to a kind of painting in which aesthetic decisions could be so systematized as to be the source of the aesthetic itself.

It sounds as if I'm making Pozzi out to be a Heisenberg-derived epistemologist. Certainly something of the case is true, but at the same time far off course. Pozzi's work insofar as it has been a source of controversy was taken to be a function of reductivist sensibility marked by a personal color and small talismanic organization.

During the early '70s when Pozzi unrepentantly resumed a painting that had paralleled his other work all the time, he was taken to be part of a vague coalition of quasi-unreconstructed painters who, after having been told that painting was dead, continued, for all that, to paint, admittedly in a flat, self-effacing, quietistic way. Several artists belonged to this group, insofar as it ever was a group. Michael Goldberg, the elder figure of this loose coalition, talked about the group as a group, but Pozzi, having experienced first-hand in Italy the twisting to self-service of Marxist idealist architectural theory, felt the incipient move to institutionalize was coercive. Pozzi is a Marxist but not a populist. Not that he would not contribute to a public awareness of this new, scarcely viable position. Writing for the Italian art journal, *Bolaffi Arte* (under the pseudonym Peter Licht; Peter the stone, Licht the light — the reversal of his initials), Pozzi contributed three important, lavishly illustrated articles during a short career as a critic. The articles, published over a three-year period, 1972-75 addressed the three main trends that Pozzi thought the most emblematic directions to have emerged in painting following the ructions of the '60s: the advent of Photo-Realism, the flat painters (who largely formed the circle of his friends and closer acquaintances), and a last one dealing with certain eccentric, non-flat painters with whom he was also associating. In a sense, these articles were written from a position of self-defense, insofar as they went against the still prevailing notion of painting's demise as incipient or as a *fait accompli.*

Among the figures of the amalgam were Ron Gorchov, Pozzi himself, Michael Goldberg, Paul Mogenson, a smattering of the younger painters from the Klaus Kertess stable — Lynn Umlauf, Judy Rifka, Doug Sanderson, Susanna Tanger. Since the exact nature of the correlations between the members remain irretrievable, except insofar as they believed in painting, however vestigial a painting it may be — one really can't see them fully as a distinct stylistic group, certainly without stretching credulity. At least some of them were drawn to the elementary logic and dualistic structures of the axioms of Bochner and Rockburne; but they pondered why these propositions had to be applied at the expense of the hand. (Indeed in the end they weren't.) This vague sense of community led to the opening of links with a broad set of artists bridging out from a scarcely definable core. Pozzi, showing at Bykert, was introduced to Paul Mogenson and Eve Sonneman (the photographer). Lynn Umlauf, then supplementing her income through cooking in the burgeoning bars and restaurants of Soho, drew in the younger figures of the 112 Greene Street Co-operative — Richard Nonas, Jene Highstein, Suzanne Harris, and Gordon Matta-Clark of the Anarchitecture group. There were others too: Denise Green, Jim Bishop, Joel Shapiro, Marcia Hafif, Jeremy Gilbert-Rolfe. For sure I miss some. In short, taking into account the broad network of Soho gallery and social affiliations, Pozzi found himself, for better or for worse, at the center of what was taken to be a nexus of sensibility-oriented epistemologists — or, if you like, of a yet more recent school of information-oriented abstract painters, whose work was marked by the modest scale of Tuttle's personalistic ineffable art. By 1975, all of this opened up, boomed as it were, and it is scarcely possible, even at this short remove, to imagine that so divergent a group of painters, sculptors, and even photographers might have thought themselves to have formed even a transient coalition. "Look Robert, the sequence is so," Pozzi says to me about his work, pushing his closed hands forward across the table top. "Until 1975 it looked like this and then, after 1975," — opening all his fingers — "it looks like this."

For all the eccentricity of Pozzi career, there is an artist who might serve as emblem — Giacomo Balla (1871-1958). Like Pozzi, Balla's career also seems a suite of brilliant disconnections, even more surprising for its antedating World War I: *The Stairway of Farewells, The Young Girl Running on a Balcony, The Path of Swifts,* the full abstraction of the *Iridescent Interpenetrations.* Naturally, a cohering thread runs through all of this, the search for a Futurist

Sign of Energy in an Environment. But, taken as a whole, Balla's work appears to lack an organic growth toward consistent manual felicity (often misconstrued as the touchstone of art). Instead, Balla offers the constant thread of style as inconsistency. Pozzi's art is also marked by his horror of formula, a distaste for what he calls the "bureaucratic" character of style manifested through "predictability" and/or "precodification." Relentlessly William James-like, Pozzi deducts empirically from experience, rather than from any larger metaphysical or abstract ambitions. Despite this, Pozzi admits that this kind of empiricism also invokes "an order of approach, that of change even, which is itself capable of becoming its own kind of norm." Such "norms" signal, for Pozzi, moments of "cultural alert" because they so easily play into the demands and appetites of the publicity media. In reacting against these "norms," fresh artistic moves are postulated — else the artist achieves no more than the shoring up of a merely normative or frozen style. The norm then may be seen as a kind of rule that in turn indicates the tool, a methodological mechanism or a path of discovery. In acting on the fresh impulse provoked by the norm, the utopian dream of public and private reintegration once more may be affirmed in the artist's work: "With rules come the state of emergency; with tools come the state of discovery."

Still, intentionality — even when expressed as an intention to reject style — is no guarantee of art, though in the end it may be all there is to Pozzi's art. Said simply, Pozzi views art making as "just doing," granting the continuing examination of the basic experience of the visuality we call painting. But we misapprehend Pozzi when we do not perceive the links binding Pozzi's art episodes together — however distant one component may be from the others. These bonds may take the form of

> . . . imitation; force of gravity; the incorporation of accidents and context; dualisms, simple or multiplied; adding; removing; translations or transpositions from one medium to another, the six basic colors of painting mixed or alone; the arranging of information in clusters and series (progressive or constant); patterns, either regular (grids) or irregular (textures); and an awareness that any elements can be combined so that density,

distance and contrast, both within and without, are detectable and describable. These mechanisms are the base from which fantasy, intuition and experimentation can start off in unpredictable directions.

(Lucio Pozzi, "Letter to Peder Bonnier," 28 July 1979; edited by Peter Zabielskis)

Pozzi's definitions of painting are exemplary for an absence of metaphysical posturing; for example, "color transforms something which is something else and makes it look different"; or, "Painting is something that makes sense but is not logical." Of course, the absence of expressed intention may be, in its most extenuated sense, a form of intentionality leading to no error, no mistake, no slip-up, everything as it happens, hence on purpose.

In the degree that such a philosophical premise is even possible, Pozzi may be said to embrace, as Balla had before him, unintentionality as a model. Yet to call upon historical models or artistic paradigms abrades Pozzi's naunced views. Rather, he seeks reconciliation with a certain strand of indecision and ambiguity in Modernist history. For Pozzi, the drift toward sheer ambiguity in art-making begins in the first assemblages, in the initial pictorial-sculptural conflation that took place in Picasso's painted constructions of 1912-1914. This eposide is uniquely heroic for Pozzi. The Synthetic Cubist assemblage — in addition to what it does do —also stresses elements that escape artistic control and volition. This places art-making in a position open to "deconditioning forces" or "uncontrolled properties." Thus, it is easy to grasp Pozzi's view that the Synthetic Cubist assemblage generated, in large part, the automatism of Dada and Surrealism. This "other tradition" of 20th-century art is the one to which Pozzi feels drawn, though never in an iconographic sense, only in a process sense. This evolution, strung out to the present, goes far in explaining Pozzi's identification with the elaborate paradoxes of John Cage and Jasper Johns.

This last development toward the third avenue — that of Johns, Cage, Rauschenberg, Cunningham, etc. — is far more culturally linked in Pozzi's case than might be expected. Pozzi arrived in the United States as both European and pilgrim. As the one, he is marked by a liberating nostalgia for something given up, eschewed — and in being given up, being liberated; as the other, that is, as pilgrim, Pozzi actively, participationally, sought the new condition of exile — one that studiously rejected an inquisitory Marxist stance — especially that of a declaration of intention enunciated prior to invention. Not only did Pozzi's ingrained Marxism fall away from him here, but so too did the Futurist-Marxist model of perpetual revolution — since, for example, the Futurist manifestoes jumping the gun on their superannuation had become for Pozzi merely a new set of encyclicals.

Four culprits of rationality stand indicted in Pozzi's fine-slicing shredder: Modern novelty, Romantic object making, the Renaissance commodity fetish, and, in the grandest indictment of all, the broadest plaits of European rationality, be they Socratic, Aristotelian, or Platonic. All this was shucked off in the name of another rationality perceived by Pozzi, one that seeks a more authentic level of visuality, form in its more pristine expression (Plate 38).

This, of course, brings one full cycle, to those very qualities that mark Pozzi's work as being so thoroughly "European": its richness of contradiction and incongruity, its Mannerism and the concomitant sense of elegance germane to mannerism. Elegance implies, more than anything else, an awareness of the contradictions and paradox inhering to all human activity. Still, though the sensiblity may be European, its self-evident pragmatism is utterly rooted in the traditions of American philosophy — in short, the work could only have happened here, in New York, at a certain compelling moment of recent New York history.

Of course, it is impossible finally to say how or why it is that artists might find themselves in such a position. Little trite phrases disguise or minimize Pozzi's obliquity: he "feels good" in paradox, he "feels authentic" in it. Mind you, Pozzi is marked by the heroic years of the American hegemony; and emerging in the salad days of early post-minimalism, "elegance" is a difficult term of admission for him, despite the manifest elegance of his work. Still, we can stay with Pozzi's evasive conversational gambits a bit longer, since the sentiments of "feeling good" and "authenticity" in Pozzi's usage are liable of yet greater repercussion. Such existential clichés in his terms imply an historical assessment. The artist — Pozzi really — is "a fine instrument in the culture who disallows the norms that play into the media." From 1965 on, then, one sees how clearly Pozzi evaded the pre-digested cliché insofar as there appears to be no formal consistency locateable to a coherent methodology. Yet the method for all that remains consistent. So too does the infinity of corollaries — Pozzi's art, that is — that forms the fallout of this syllogism.

# ENTRIES: OEDIPUS RECONCILED

In the last years of the past decade, Richard Serra forded two sculptural extremes. One was a blunt, unapologetic investigation of industrial matériel, an orientation akin, at the time, to the work of Carl Andre. This mode most clearly expressed Serra's continuity with Minimalist sensibility. By contrast Serra was also drawn to an expressive gesture that affiliated him with the revival of Abstract Expressionism and a process-orientation equally marked in the period. This side of things came to be recognized as the first type of Postminimalist expression, a refreshed pictorialism in sculpture shared commonly with early Benglis, early Le Va, Eva Hesse, and others.

These extremes, so marked in early Serra, were also bound up with the profound Oedipal syndrome that marks the evolution of American style, a history in which the father is killed, so to speak, as if to acknowledge the paternity all the more. The father totem, in this case — David Smith. Serra, determined to continue the traditions of Smith, rejected Smith's essential tool, the oxyacetylene torch. This allowed the maintenance of the idiom of a Cubist-derived Constructivism while camouflaging, at least for the time, a suppressed, even closeted, commitment to Constructivism, for, in the late '60s, Constructivism had fallen into a debasement similar to the bankruptcy of latter-day formalist painting. In rejecting the torch of Constructivism's master, Serra came to rely on stresses, pressures, masses, cantilevers, balances, and other natural forces and weights to create a cool, crypto-Contructivist work that encoded simple verb infinitives — to rest, to cut, to lean, to intersect, etc. These direct postulates, of course, could take on more expressive intonations, especially, say, when Serra chose to toss ladlefuls of hot lead or to distribute torn pieces of lead sheeting as functions of his process-related concerns.

As such, Serra seemed a central sculptor of Postminimalism. But one of the truths of the history of style, especially in the twentieth century, is how quickly it is depleted. So while many were drawn to the style that Serra so forcefully inaugurated, he himself was leaving it, reverting back to a Constructivist mode — as Bochner, say, also returned to painting — as early as 1971. In reviewing an exhibition in which Serra participated in that year, I noted the awakening of the Constructivism that had remained dormant the preceding four or five years.

But that Serra may have, at this moment, opted to move outside a proscriptive Postminimalism — then fast shrinking to an exact diagnostic profile — he was also answering needs organic to his own development. His was no accommodation to exterior critical notions, but a necessary response honoring internal exigencies, no matter where they would carry him. And it brought him back to Constructivism, but one of an enormously open field of interest. Not for him the small part-to-part comparison, the glorified paperweight of the confirmed David Smith knock off.

Many possibilities — at least five major ones — asserted themselves this past decade: (1) comparisons made over vast terrain-locked elements; (2) sculptures, the elements of which sparked shifting perceptions [as speculation aimed at mentally imagining (or participating with) components only but hinted at through the use of partial, even fragmentary evidence]; (3) enormous, often geographic scale. In all this, (4) Serra bracketed Malevich to Newman, thereby allowing the Suprematist graphic motif to assert itself even more greatly. This may be an additional clue to understanding the close connections in Serra's work between sculpture and the architectural or site-specific drawing. In a certain sense, vast sprawling works continue the romantic and pictorial side of Serra's mind.

Then too there was (5) the even greater compaction of industrial substances, more densely compressed molecular structures of steel plates and cubes forged under technological weights hitherto unexamined in sculpture. Such properties can be said to continue the classicizing Minimalist side of the sculptor's sensibility. Even if the '70s saw Serra's return to the Constructivist mainstream, his

version of Constructivism modified and enlarged what the term means — and for precisely those mental and physical qualities from which early on he had distanced himself.
It is egregious to rebroadcast "Slow Information" (1969), an old essay on Serra, but as I glance through my private journals, many stray entries precisely refer to those new Constructivist developments that Serra disclosed this past decade: Fast Information.

*17 January 1971*

Philip Lieder [then editor of *Artforum* magazine] and I going to the Bronx — 183rd Street and Webster Avenue — to see Richard Serra's piece *To Encircle Base Plate Hexagram,* as part of, but not included in, the museum precincts of this year's Whitney Annual. The work is implanted in an asphalt street that slopes up to a heavy stone palisade from the height of which one looks down on the work. This piece is perhaps the most understated kind of monument possible to a neighborhood that resists, it is maintained, the notion of art as high abstraction. Lining the street is the rubble of razed, burned-out tenements. (For a second I thought that certain mounds of brick or broken concrete were meant to be the disingenuous solution to the problem.)
The work designates a flat circular site. It's made of steel, laid like a collar into the asphalt, in the figure of a circle. Half the circle has a wide flange, the other narrow — options open to the facings of a 90° corner-beam forged into a semi-circle. From the overlook, the work suggests that once there may have been something there — a trolley-car turn-table, a water tower — but, like the tenement blocks that had once bordered the site, no longer extant.
There's something about this ringed figure that is both religious and ambiguous. Phil said that when Serra and Carl Andre were in Japan, they were struck by curious structures, pile or mound-like erections of some kind, placed at the ends of little streets. The sense of these structures was neither clearly intentional nor were they accidental or impermanent, like rubble at a building site. And they liked them.
My deepest association concerning Serra's circle is that of Brancusi's *Table of Silence,* but somehow sunk to ground level. Phil thought

24

24. Mel Bochner, Apsides, 1977. *Casein on wall, 101¼" x 62¾".*

25

26

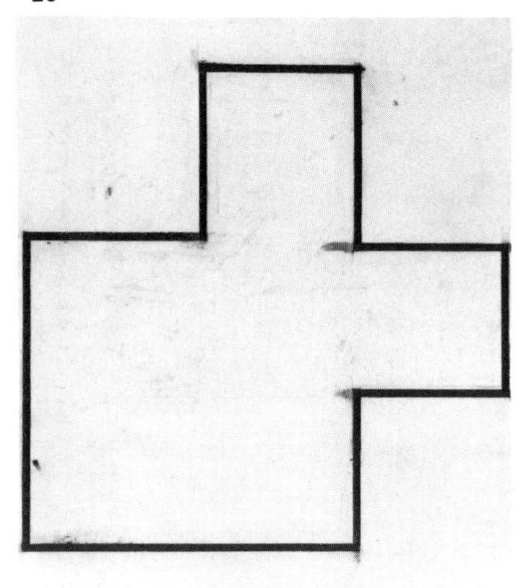

25. Joel Shapiro, Untitled, 1972. *Bridge, balsa wood, 3" x 20¼" 3"; Boat, balsa wood, 1⅝" x 11⅝" x 2⅝"; Coffin, balsa wood, 1¾" x 7¹/₁₆" x 2¾"; Bird, bronze, 1¾" x 3¾" x 2¾".*

26. Joel Shapiro, Untitled, 1976-77. *Charcoal on paper, 19¾" x 25".*

27

28

27. Sol LeWitt, Wall Structure (Diamond Shape), 1977. *Painted Wood, 7′ 33″ x 13′ 6″ x 1″.*

28. Scott Burton, Granite Chairs, 1981. *Granite, 30″ x 36½″ x 40″.*

29

29. Scott Burton, Table from Four Tables, 1978. *Bronze, 42″ x 27″.*

30

31

30. Benni Efrat, Putney Bridge, 1976. *Film and performance.*
31. Siah Armajani, Fifth Bridge, 1979. *Wood, Steel and stain, 12′ x 90′ x 36′.*

33

32

32. Lucio Pozzi, Four Windows (detail), 1977. *Installation at P. S. 1: Sheetrock, wood, four chairs, paint, 12′ x 28′ x 12′.*
33. Lucio Pozzi, Four Windows (detail), 1977.

34

35

34. Giorgio de Chirico, Furniture in the Desert, 1926. *Oil on canvas, 32″ x 39½″.*

35. Lucio Pozzi, Moon's Hook, 1967. *Acrylic on canvas, 77½″ x 64″.*

36

37

36. Lucio Pozzi, Shootout #1 (Execution), 1971. *Performance, Milan.*
37. Lucio Pozzi, 29 Models Descending an Art School Staircase, 1971. *Performance, New York.*

38

38. Lucio Pozzi, The Migration, 1981. *Oil on canvas, 104″ x 108″.*

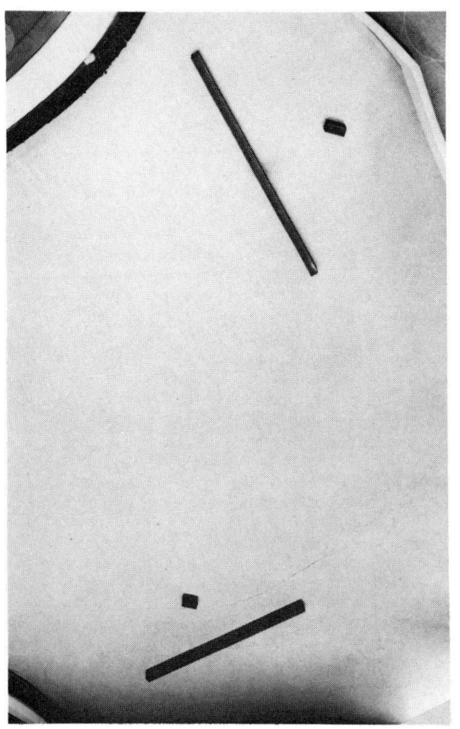

39

40

39. Richard Serra, Different and Different Again, 1973. *Hot rolled steel (4 pieces), two: 1′ x 15′ x 6″; two: 1′ x 1′2″ x 6″.*

40. Richard Serra, Shift, 1970-72. *6 sections, total dimensions, 815′.*

41

41. Richard Serra, T. W. U., 1980. *Corten Steel (3 plates), each 36′ x 12′ x 2¾″.*

42

43

42. Jackie Winsor, Small Circle, 1969. *Hemp, 8½" x 39" x 39".*
43. Alberto Giacometti, Suspended Ball, 1930-31. *Plaster with metal, 24".*

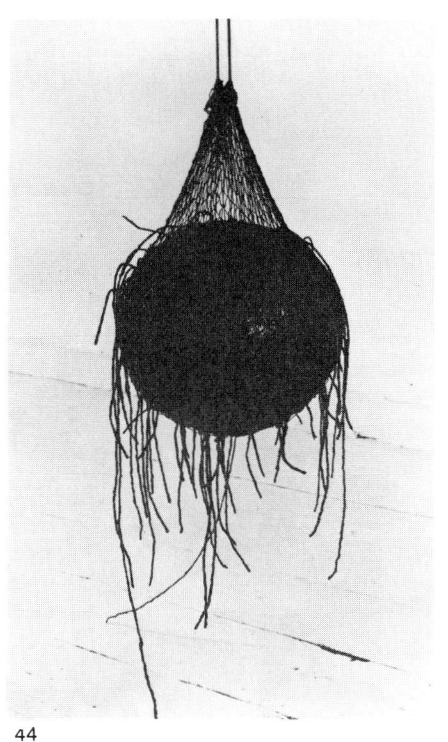

44

45

44. Eva Hesse, Vertiginous Detour, 1966. *Painted rope, net, and plaster, 23".*
45. Jackie Ferrara, The Ball and Saucer, 1972. *Paper and rag pulp. Destroyed.*

46

47

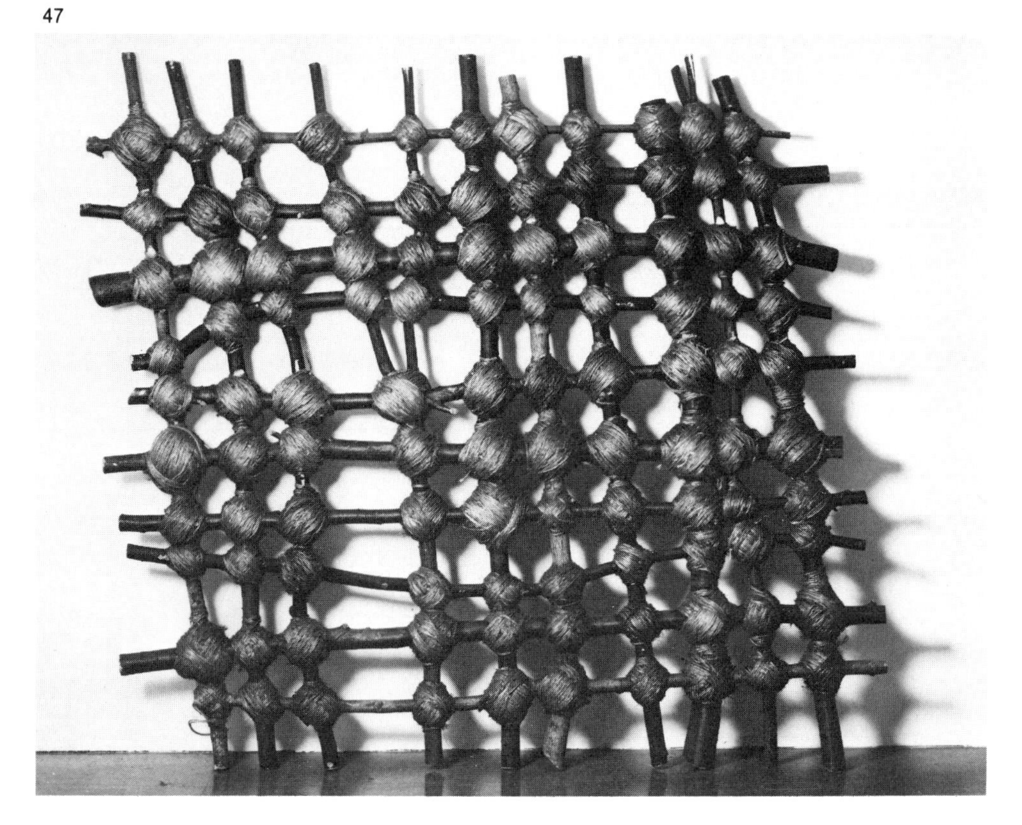

46. Jackie Winsor, Nail Piece, 1970. *Wood and nails, 7" x 82" x 8".*
47. Jackie Winsor, Bound Grid, 1971-72. *Wood and twine, 84" x 84" x 8".*

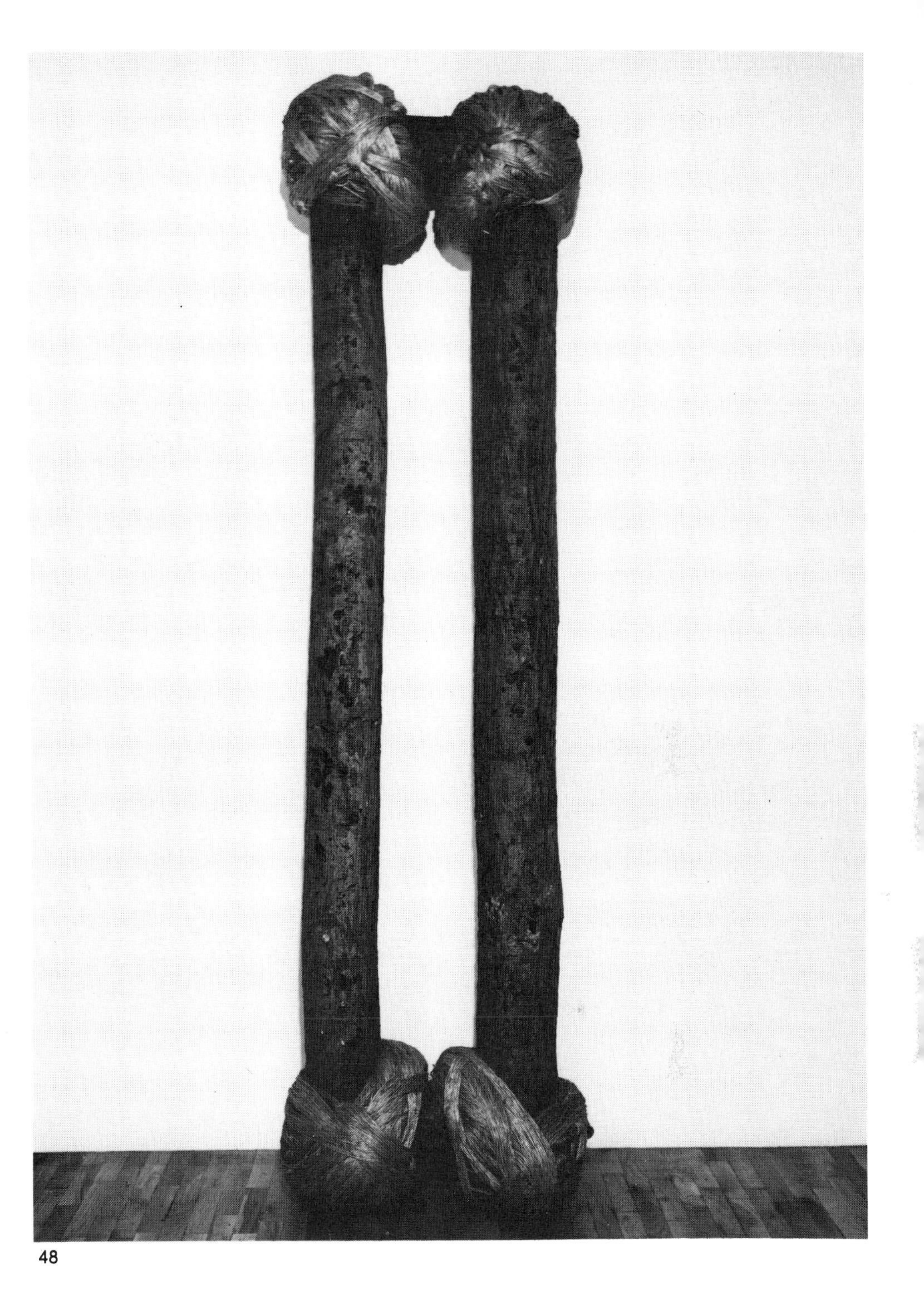

48

48. Jackie Winsor, Bound Logs, 1972-73. *Wood and hemp, 108" x 31" x 18½".*

49

50

49. Michael Hurson, Thurman Buzzard's Apartment, 1973-74. *Balsa wood, 8 5/8" x 69 1/2" x 25 3/4".*

50. Michael Hurson, Edward and Otto Pfaff, 1974-75. *Oil and silkscreen on canvas, 36" x 96".*

that the changed flange width of each semi-circle was akin to yin and yang.

*10 March 1973*

Went to a Serra talk to students held at the Whitney Museum that touched on numerous interesting details. The visit to Japan made a great impression on him. He says he now wants to go to Machu Pichu. (Is his father from Peru?) Japan meant learning to look at things for great lengths of time. When puzzled by something, he always asks "What's that all about?" in a piping voice. From the audience, I inquired after the Bronx circle in the ground. He wants it to be looked at the way the Japanese regard their Buddhist gardens. He spoke of a solid, rather tangible conception of sculpture at odds with the disintegrated, expressionistic lead tossings. Sculpture for him has come to reassert a formal, emotional, psychological meaning — though only the formal can be talked about. He's still interested in Robert Smithson [who was to die tragically four months later], Carl Andre "when he works," and Mel Bochner, "who is, you know, a sculptor."

*4 July 1973*

I saw the beginning of the actual wall painting of Bochner at the Sonnabend Gallery and remarked that the wall is too long, tending toward a kind of Mannerist distortion of a rectangular space. Yellow rectangles appear to float in the white atmosphere with red and blue rectangles yet to be added. Something about this "control of space" reminds me of the dense metal cubes and extended oblongs [called *Different and Different Again, 1973*] which establish parallel formations (one has to imagine them) in Serra's recent large, shifting installations at the Guggenheim Museum (Plate 39). The parallel bars and cubes of Serra's recent work at the Whitney annual [*Elevation,* 1973] also came to mind, especially as they were widely displaced, elements being set on the floor, distant from one another and quite up close to the wall with a broad plane of "empty space" in between.

*4 June 1974*

Saw Richard Serra today by chance. He described a new project for Cologne He attributed the source for the 120-foot sited wall to his recent experience in Peru. The project is an enclosure dealing with perceptual discrepancies occasioned by positioning oneself within an oddly formed four-sided unit. There are two 6-foot entries and several 9-inch breaks in a wall of Corten steel. From certain positions the wide entrances read as thin as the narrow breaks in the wall. Because of the discrepant relationships between the wall sections, persons standing within the piece, equidistant from one another, appear disproportionately larger or smaller in an exaggerated perspectival illusion. Also, though enclosed, the view of the city through the narrow slits is quite complete — a bit like seeing it through snow goggles. He plans to call his work after a Peruvian site.

He at first submitted a project that the city of Cologne is more greatly keen to have. But Serra rightly does not want to do it, as it is very much in line with the logarithmic expansions perhaps too reminiscent of Smithson's early work. [I'm unsure as to the fate of these projects, but aspects of them are maintained in the *Untitled Piece for Bochum* (1977), the tentative project for the Ohio State University (1978-79), as well as the trepanned Cubism of *T. W. U.* (1980), set up on West Broadway between Franklin and Leonard Streets.] Serra, for all his single-mindedness (probably because of it), is perhaps the most "artist artist" I know. Anyway, it's a toss-up between Mel and Richard.

*4 June 1978*

To Richard Serra's King City, Canada piece just beyond York University in a bucolic field — clouds forming, a jet plane's vapor trail marking our arrival. *Shift* (1970-72), (Plate 40), as it is called, is naturally a disappointment — as would be expected — on seeing so legendary and yet so isolated a work. The first effect was to "think deep thoughts." In the rising warmth of the day, swarms of mosquitoes were aroused by our walking through the already high grass drying in the morning sun. On investigation the piece grows

beautiful. There's something here of Richard Long become a Don Judd or an Andre. This severe work allows itself one or two aestheticized flourishes. Acute cuts mark the top of the wedge-driven walls, postulating a vague parallel memory of some specific enclosure.
I think of wedge walls in a continuum of wedge shapes from Morris to Judd to the *Seed Bed* of Vito Acconci, even Duchamp's *Wedge of Chastity* — though I know that Serra is disinterested in the historical iconography of abstract elements, but rather seeks an enormous physical investigation of, here, more than 1600 feet one way, and that doesn't include the valley between the slopes. Still, iconography determines form — at least in some measure — and it is possible, even reasonable, in the late twentieth century to speak of an "abstract iconography."

*7 August 1979*

Late in the torpid afternoon I met Richard Serra. He is setting up a new piece. At the end of Canal Street there is the Holland Tunnel approach, a wide, swerving island approachable on foot only across an industrial overpass. This traffic island, covered in gravel and a bit of neglected landscaping, falls under the jurisdiction of the Port Authority. Serra got to the agency, said he wanted to site something on the island, and received a go-ahead signal. At the moment there are 12-foot-high poles tipped in yellow, forming an almost perfect quadrant "eyeballed" into the gravel, 90° of a circle that would be, if complete, some 800 feet in diameter. The wall intended to occupy the demarcated curve will be made of Corten steel. One sees how the monumental wall is a function of the curve of driven access. For the motorist, the wall shifts in an ambiguous perspective from the condition of line (that is, vertical edge) to a long curving plane that defies easy, perspectival comprehension.
Richard has studied the site from the windows of the surrounding buildings; he has even seen it from the air in a helicopter. For me, the thing works only in blueprint, as an elegant linear curve. Once the piece is identified as wall, especially steel wall (open to graffiti attack, behind which derelicts live), the project will not be supported. But it is all so typical of Serra; coveting the site, he

nonetheless avoids an acceptable solution, that is, a vertical construction which could be seen as a kind of an insignia or emblem, or beacon. I don't want him to do this, but should he, such a proposal would handily succeed. I want him to do what he wants, but what he wants won't pass — not because it's not good or important, but because it will be shouted down owing to sociological anxieties.

Built into Richard's dedication to sculpture is a kind of set-up situation: not to fail (for he cannot fail), but rather one of immensely strained resistance, leading to a heightened sense of failure, or rejection, of grand effort put forth without the concomitant approval he also hopes for.

His hair cropped short is now grizzled. His stong body has grown thickened and squared-off, so that he seems older than he is.

*27 April 1980*

On Thursday last, a reception for Richard Serra whose great sculptures are now part of New York. The *St. John's Rotary Arc* outside the Holland Tunnel is the curved, horizontal, low-lying counterpart to his vertical piece *T.W.U.,* downtown between Franklin and Leonard Streets (Plate 41). *T.W.U.,* in terms of its part-to-part open Cubist Constructivism, symbolizes sculpture itself. The broad face of the *Rotary Arc* symbolizes painting. One could be even more specific: the very edges of the arc enact the "zips" of Newman. The paradigms for these two works are Brancusi's *Table of Silence* and the *Endless Column* at Tirgu Jiu. Sidney Geist once pointed out that the urban monuments that inspired Brancusi at Tirju Jiu were the *Arc du Caroussel* and the Obelisk in the Place de la Concorde. Accepting these connections, Obelisk = *Endless Column* = T.W.U. and *Arc du Caroussel* = *Table of Silence* = *Rotary Arc,* further enlarges the historical ironies implicit in public sculpture. The distance separating *T.W.U.* from the *St. John's Rotary Arc* to the south marks, to the north, its distance too from a dead triviality, a recently erected and inadvertently droll effigy of Juan Pablo Duarte (by a certain Nicola Arrighini). It's as if, Serra's pieces argue the symbolic status of painting and sculpture between themselves and the sculptor's relationship to Brancusi, while they also argue against the dead representational traditions of public sculpture by which we've been mostly abused these last few years, viz., the ghastly *Alice in Wonderland* in Central Park and the horrendous Einstein Memorial in Washington, D.C.

*4 May 1980*

Stroll up from City Hall and pass *T. W. U.* Graffiti commences — a boldly crayoned "Fuck Art"; and someone has tossed a bucket of red paint high on the work so it could dribble down. So it goes.

# WINSOR KNOTS: THE SCULPTURE OF JACKIE WINSOR

I remember asking Paula Cooper, Jackie Winsor's dealer, just what, after all, was the meaning of Winsor's sculpture, that is, what meaning did her work have beyond a merely functional description of the individual pieces. Cooper supplied the following précis: "They are well considered entities that take up their own space." This seemingly scanted reply capped a long discussion of the artist's work, and is exactly what Winsor's sculptures are about. Cooper was not being flip. Her reply, based on an appreciative familiarity with Winsor's work, is very much in keeping with the way the artist herself talks about the work: the desire to shape the exactly right reply, to find a nicety of meaning, often forces the response to take on the coloration of an evasion.

Winsor's background goes far in allowing us to gauge some of the sources of this reluctance to translate formal experience and sculptural practice into the possible misrepresentation of words.

Winsor was born in Newfoundland in 1941, to a family of some three hundred years worth of Newfoundlanders. Thus, she was born, so to speak, in the North American eighteenth century, in a place marked by separatist leanings and a quite literal insularity.

Only during Winsor's childhood, in 1949, was Newfoundland made a province of Canada. Agnes Martin, Dorothea Rockburne, Ronald Bladen, George Trakas, and Michael Snow, among others, are also Canadians by birth. This circumstantial detail provides Winsor with a natural curiosity about their work in ways having little to do with formal similarities or programmatic affinities. Winsor's attitude towards her colleagues, whatever their national origin, is largely one of indifference: they are there, taken for granted, a communal bedrock.

Winsor's father's people had been constables and ship captains; her mother's, farmers — comparative newcomers to Newfoundland, there but some two hundred years. In 1951, Winsor's father, a factory foreman with a natural bent for engineering, emigrated south to Boston — Cambridge really — hoping to find greater opportunities for his three daughters. Boston represented a radical contrast to Winsor's Newfoundland home at an age when such distinctions were most sharply experienced. The country suddenly became the city. Though not farmers, the Winsors were countrified. Indoor plumbing was a novelty. Jackie had never seen a black. The onset of puberty had begun. In Canada, Jackie had been a tomgirl. Despite her family's liberal Protestantism, no notion that one might want to be an artist prevailed. As a possibility it simply never had manifested itself. "Professional woman means being a nurse, at best a teacher. Women had broods. there was just nothing for women professionally."

So the family came south. Jackie was taunted for her Queen's English by Boston schoolchildren whose regional accent is among the most pronounced in America. "I couldn't stand it here. I thought it vulgar. Until I was eighteen I returned to Newfoundland every summer, staying with first one aunt, then another. Boston was such a shock. I reconciled myself to puberty when I was twenty. Canada was my childhood. My adolescence was here."

Winsor's high school years were spent at Cambridge High and Latin School. That Cambridge was the seat of Harvard University seemed not especially important. "I started taking advantage of Harvard in my senior year in high school. I had no intention of going to college. It had never happened in my family. I even spent an extra year in high school. As a junior I started going to art lessons at two art colleges. I had always taken art as an elective in high school. My art teachers, Mrs. Ritterbush and Mr. Santoro, sent me off to the Saturday morning art school programs designed for high school students in the Boston area, at the Massachusetts College of Art and at the Museum Art School." Winsor graduated high school in 1960 and took odd jobs — in order to put aside funds to go to the Massachusetts College of Art. Her last year in high school — tutored, working hard — was marked by a drive just to be accepted in art school.

Between 1961 and 1965, she generated her Bachelor of Fine Arts credits. "The ambience at M.C.A. was very regimented. I did a little bit of everything."

All of this presents a conventional unrelieved consistency. The art produced at this time was "kind of representational and expressionistic." Winsor as yet had no real sense of herself as an artist, although it was "all very exciting and I loved it. It was a whole new experience for me." Still, there was a teacher of color theory, Emma Lennon. Winsor doesn't know what her work was like, indeed if she made any, but "she was a woman, and she was doing something. She was peculiar. Her persona was very exciting even though she was older. She was a woman who was doing something and she was professional."

In 1964, Winsor went to the Yale Norfolk Summer School. "For the first time I met artists rather than art teachers. They argued a lot about art." Winsor never entered the lists. Al Held, Louis Finkelstein, Bernard Chaet, among others, passed through that summer. That they may have given Winsor a "crit" was far less important to the artist than an exposure to people who were "artists before anything else. From nine in the morning until ten at night people were painting, taking themselves very seriously" even if the paintings produced were sensibility-based landscapes. Winsor effected a transition out of figuration while taking up photography; photography would continue to engross her through graduate school. "It was a way of making form important — photography had a more refined grain than the very heavily manipulated, stroked surface of painting."

The critical, verbal side of the teachers and students at the Yale Norfolk Summer School failed to impress. "I couldn't identify with verbalizations. The arguing was not about intelligence but about who won. There's no reason to be right about art."

In 1965 she began her Master of Fine Arts degree at Douglass College, Rutgers University, New Jersey, working primarily with the painters Ulfert Wilke and Reggie Neil. Tellingly, Winsor, despite long years in various art schools, never studied sculpture in any formal sense.

During the run of her college years, B.F.A. through M.F.A., there had been occasional trips to New York. The choice of Douglass

College was made primarily because of its proximity to New York. Winsor, at some length aware of politics, factions, stratagems, and worlds within worlds that make up the New York art scene, has always chosen to back off from these aspects. "I'm just not political that way. My interest is in making things that are important to me because they are real, experienced, because they are tangible. I am interested in transforming an idea into something physical."

"Is there any quality in your work today that is traceable to your college times?" I asked. She replied, "Energy going into itself. solidity. Overallness. Those are qualities that come to me from those days." Perhaps it can be seen a bit like this.

Her student work was figurative and revealed a proscriptive interest in anatomy. This led to drawings in which muscularity was exaggerated to a point where later in sculpture it was seen mainly as a sequence of bumps and repetitive ridges. The overallness, the uniformity of surface, attracted the artist to rope and the process of coiling. This in turn led to work embodying a certain sense of the monumental: that a psychological state may be greater than the literal space it occupies.

Since the rope pieces grow out of the immediate environment of New York, Winsor's art recognitions traceable to that time, many of them negative, require some note. Between 1965 and 1967, when she was awarded her M.F.A., she grew aware that "not much was available to women. When I came to New York in December of 1967, my first impression held: women were transparent." Between 1967 and 1969, "I had a clear idea of my invisibility." Perhaps all this is so. It certainly corresponds to tractarian feminist history of a kind now foregone in the women's movement.

I visited Jackie Winsor in her studio in 1969, an odd triangular space overlooking the Manhattan Bridge access and the old diamond market at Canal Street and Bowery. Soho was as yet *terra incognita;* space was still cheap. Her studio was filled with the early monumental coiled rope pieces (Plate 42). I saw them in terms of the analogy of materials — rope to cord suggesting a continuum of fat to skimpy — which they presented with the work in cord by Keith Sonnier, her husband, who led me to her studio in the first place. When questioned about these contingencies, the artist bridles; it misapprehands her sense of artistic autonomy. The building of a career, of a separate identity, was so difficult for a woman artist at that time that Winsor is still reluctant to enter my sense of historical and critical affiliations, preferring instead to allow each artist the integrity of an individual achievement. In Winsor, this attitude amounts almost to a professional credo.

What of the connections to the work of Eva Hesse? The substances that Winsor worked with in 1968 — say latex and rope — engage substances and procedures much associated with Hesse's work. Winsor reflects: "The way she would finish something was exactly the way I wouldn't have finished it. I would be clearer about the form, how the surface got to be there in the first palce. How it reveals itself in the process of its making. She seemed more interested in the material itself. And now since her death she's been blown out of scale by the women's movement — deified."

While Winsor was at Douglass College, the earliest independently minded pieces appeared. That they are synthetic — amalgams of sculptural work that the artist knew — in no way denies their interest. They amalgamate issues and substances in a way that expresses an awareness of an alternative to the conventional constructivism — too often fobbed off as the only option open to modernist sculpture — epitomized in the suite of David Smith to Anthony Caro to the Bennington School of sculptors. By contrast to that academy of constructivism, Winsor's interests in 1966 address a colorism, a pictorialism, and an imagery roughly traceable to Surrealism — not Cubism — especially Surrealist sculpture of the 1930s, a source that marks the broad range of early counter-minimalist sensibility in terms of a focus on compulsive procedures and eccentric substances.

At that time Winsor was contrasting highly polished and finished abstract elements — sphere sections or rectangular solids — against organically inferential passages, so that cubes, say, seemingly sprouted hair or tails, not to mention even more anatomically specific forms, testes-like or phallus-like in character, though as always, Winsor plays down this latter aspect. "I was doing what interested me. I'm not theoretical in orientation. The forms arose out of studio practice. I came at it as a painter. You never lose that."

These sculptures were made in hand buffed clay or cast in resins and plasters. The surface results literally from the casting process; there is no subsequent polishing, although refined pigment and resin combinations can result in works of extremely delicate coloration and/or subtle sheens. [An exaggeraged sense of editing has prevented Winsor from showing these works.]
These student sculptures of melon-like sections have a prototype in the Giacomettis of the 1930s. In perhaps his most characteristic work of that early style, *The Suspended Ball* (Plate 43), a ball depends upon the ridge of a crescent shape, even to the point that a slight indentation is marked upon the sphere itself, inferring sexual roles — the crescent male, the sphere female. This tradition, maintained in Surrealist sensibility, finds a parallel in more recent work, say by Eva Hesse, such as the *Vertiginous Detour* (Plate 44) in which a sphere is caught up in a net from which strings dangle, or in the *Ball and Saucer* (Plate 45) of Jackie Ferrara, made out of paper and rag pulp or in Jackie Winsor's own *Solid Cement Sphere* (1971). I don't mean by this observation that Hesse influenced Winsor or that Winsor influenced Ferrara; but rather they all maintain an anti-constructivist, anti-formalist stance in their sculpture — at least in their early work — one that invokes sexual nuances through substance and methodology whether or not the artists were even aware of a specific model in Giacometti. This is more a question of the kinds of imaginative acts that an anti-formalist sensibility allows in contrast to the restraint typical of the formalist ideological stream. Winsor says, "These works were not intended as sexual shapes, though the sexuality may be there."
Last, the range of substance which Winsor explored at that time included sensitive rubbers, latexes, and polyester resins. These extremely pictorial and painterly substances, typical of the early phase of Post-Minimalism, have been abandoned in her work since 1968, though the fibrous or hairy textures of rope as well as its surrogate linearity continues to refer to the equally pictorial and painterly predispositions that she brings to sculpture, ones that first received definition in the mid-sixties. It is surely not circumstantial that the more pictorial and eccentric range of substances she exploited at that time found echo in the latex, string, and rag pieces

of Keith Sonnier's earliest phase. But always, Winsor's work must be differentiated as it emphasizes ritual and duration, radically at odds with Sonnier's extroverted expressionism.

While the methodologies she employs — hand based, repetitive, exhaustive — may have been traceable to non-constructivist sculpture, that is, Surrealist sculpture of the 1930's, and although her substances and methods may reveal a crypto-sexual tagging, Winsor's essential forms nevertheless derive from the icons of Minimalism. The square, circle, and triangle, the last being more rare in her work, are the essential figures of a prevailing contemporary sculptural/architectonic style. In part, the very abstract reductivism of Minimalism (c. 1962-1966) triggered, by its sheer austerity, a widespread pictorial and expressionist response in sculpture of the later 1960s, notably, though hardly exclusively, in the work of women.

Of course these circles and squares had been given ideological preeminence for the American Minimalists — Judd, Tony Smith, early Stella, early Morris — in their having been sanctioned by Malevich for whom they formed the basic Suprematist elements (c. 1915), and more surprisingly by Kandinsky. In *Concerning the Spiritual in Art* (1912), Kandinsky had first codified this chart of primary shapes. At length the near-mystical belief in the inviolability of these forms can be understood as an extenuated Neoplatonism, for throughout the *Discourse*s and the *Epistles* Plato indentified the perfection of the sphere and cube as beautiful in themselves, but in no sense in any relative way: "What I mean . . . is something straight, or round, and the surfaces and solids which a lathe, or a carpenter's rule and square, produces from the straight and round . . . Things like that, I maintain, are beautiful not, like most things, in a relative sense; they are always beautiful in their very nature . . ." *(Philebus)*. Not that such shapes are free of an absolute meaning for Winsor; rather the exact nature of their specialness is hard to pinpoint. It appears that such figures and forms in Winsor's mind "mark off a private space." It's an odd idea since in a certain sense, squares and circles are, by the very success of abstraction in modern art, perhaps our most public artistic designations: "It ends up not being the shape per se, but what activity the shape is involved in."

From 1969 on, it became clear that with rare exception Winsor's sculpture would be based on these Neoplatonic referents. Thus, the covert content of Winsor's sculpture is located in method and process as compared to the outer form of the work — which is overt, a cube or a sphere, a kind of known and received fact. Variations in scale as they occur do so from Winsor's need to negotiate generalized forms that duplicate themselves easily in the mind with a methodology in which the hand, the body, and rudimentary carpenter's tools — hammers, chisels, clamps — play a primary role. Thus, the work takes on a warm, affective character. Its scale neither exceeds nor shrinks from a basic human module — the arms outstretched, the arc of the descending forearm, a kind of handlebar length of rope. This description of the general intention of Winsor's work emphasizes the separation between form, substance, and process. By contrast, it is Winsor's contention that work identifies the congruency of process, hand and shape. "My intent is to make them seem together, integrated."

The first wholly mature works were the rope pieces — the *Untitled Rope Cyliner,* the *Double Rope Column,* at length the *Rope Circles* and *Double Circles.* These appeared in 1969-70. The method explored processes of rope twisting and invented rope technologies, so that the diameters and heights of the sculpture were based on trial and error. "Real knots never worked. There seemed no reason for them and they didn't work out very well." The slow trance-like process of exhaustive coiling or binding or winding (then as now) is a kind of physical gratification. "I just never feel comfortable unless it takes some time to do. My interest is in slowing everything down, as slow as it can be, so it's imperceivable. To create a one-to-one relationship between 'making time' and 'perceiving time' so that the form grows out of process. Process with alternations leads to different forms. There's a lot of choice."

Of course the circle's associative power, particularly at that moment of acute feminist activism, cannot be overvalued as it invokes an imagery of breast and orifice, of mouth, the navel, the vagina. For example, in the sanctuary of Delphi, the ancient Greeks marked the navel of the world, the place where the world physically connected to the universe, with their celebrated monument, the *omphalos.* The agrarian cycle of the world — guaranteeing the sequence of seasons — is assured by dancing around the maypole, a circle dance enacting the vagina, the maypole, its fecundating penetration by the phallus.

These symbologies and associations are almost infinite. However, a new emphasis on this kind of imagistic consciousness — contemporary with the provocations of the women's movement — recharacterized the intentions of artistic methodology pertinent to that moment. For example, painting and sculpture till then had been conventionally viewed as "male" activities. Sewing, skeining (a preliminary step to knitting), knitting itself, dyeing, work in soft thread-like materials were crypto-sexistically tagged as "female." This awareness — right or wrong — provoked two schools, one which held that a feminist art must emphasize such methodologies and imageries historically or conventionally thought to be female — the so-called school of "vaginal imagery"; the other faction adopted an adversary stance — it was precisely such activities and imageries as had been identified as "womanly" that must be avoided if only to break down the artificial stigma attached to the term "woman artist" and at length to claim the turf of painting and sculpture for women as well. Both views, incidently, are perhaps by now no longer the burning issues they once were — the arguments seem dated in their very writing.

Though Winsor emerged at a moment of great feminist pressure in the art world — a pressure annexing such methodological and imagistic inferences as her work invokes — she has never felt herself to be in any exclusive sense a feminist; rather her concerns are for a recognition of the status of the artist as the embodiment of something special and of merit compared to the unconscious contempt in which artists are held by bourgeois society. Thus, often at considerable sacrifice, Winsor insists upon using the Art Transfer Agreement, at the time of a sale, in which the right of artists to value accruing to their work after it has been acquired by collectors or institutions is maintained.

In Winsor's early phase, certain variations growing out of her making process appeared. In the *Chunk Piece* (1970), yard-long pieces of hemp rope have been cut and bound at the edge, frayed ends doubling back over the binding. This cylindrical formula in certain ways corresponds to the *Solid Lattice* (1970), in which lath-like lengths of wood have been nailed in consecutive sequence like a fasces around a core. This hammering sequence by which the form reveals the process — an attitude contemporaneously explored in Richard Serra's lead tossings (through his process emphasized Abstract-Expressionist speed and asymmetry — is most explicit in the *Nail Piece* (Plate 46). Here a sequence of boards is nailed one upon the other so that each layer is understood in terms of its lamination to the previous one. At length, nine layers later, the floorbound oblong (a late sixties archetype) has gained an enormous weight of nails — "the energy going into the piece is amplified by the previous energy." This may be one of the few works allowable to a specific autobiographical referent. When interviewed by Lucy Lippard, Winsor recalled the moment when as a girl her father let her partake in a building chore. Although requiring no more than a few nails, Winsor, aged nine, hammered twelve pounds worth of nails into the project. (*Artforum*, Feb. 1974, p. 56.)

By 1971, the focus on exhaustive solipsistic process — though never to be given up and in the most recent work intensely restressed — was deflected by alternative issues. In this shift Winsor was keeping pace with the general de-emphasis of process which marked progressive sculpture at that time. By 1971, for example, Serra also abandoned the painterly expressionism of his serialized lead tossings for a scale-oriented constructivism informed by Barnett Newman's late paintings. Similarly, in 1970, the last year of her life, Eva Hesse had abandoned the crafts-related alterations of open cubes and grids for a sense of monumentality derived from the gestural side of Abstract Expressionism.

By the early seventies then, a certain dissatisfaction with an emphatic process fixation in Winsor's work led to the introduction of new organic elements — saplings, tree branches, logs — fastened together by notches disguised by bound rope knots — the "Trees." The sequence moved from an organic grid - the *Bound Grid* (Plate 47) in which at one symptomatic point the crotch of a sapling forks the grid — to the individual square of the grid (*Bound Square,* 1972); and from there to the more eccentric compressed side by side format of *Bound Logs* (Plate 48) that recalls the earlier rope *Double Column.* In part, the use of the grid still relates to the Minimalist icons of pure information. No less than a circle, square or triangle, the grid is an index of rationalist structure; one that additionally implicates a temporal increment occasioned by the side-to-side or up-and-down movements necessary to read a grid. Similarly, much painting at this time — and Winsor's sculpture never revoked its painterly biases — sought to evince its expressive status by contrasting a free gesture and sensibility-based color against the stable configuration of the grid. In short, Winsor's work employing tree elements may be said to bind up sensibility and stability while reflecting, however unintentionally, a disabused society's new fascination with ecology, however short-lived this social awareness may have been. The "Trees" emerge out of Winsor's search for other materials that invoke the homey and secure affectivity of rope and twine — materials such as wood, worn brick, and certain kinds of trees, branches, twigs, and barks.

Caught up in this ambiguous associative conception of naturalness, the exaggeration of the knot in the "Trees" took on a double role, one actually constructivist — it literally held the logs together — while it equalized or balanced the extremely powerful effects of the wood texture and substance itself.

The twining, skeining experience, as before, continued to gratify. The endless repetitive character provoked in the artist a constant attention and fresh decision making: add more rope here, make sure the rope hasn't slipped there, fill in and negotiate the general profile, etc. So compelling did this carding become (certain of these pieces required the use of 600 pounds of twine) that one could say that *Four Corners* (1972) — a square frame each corner of which was so exaggeratedly bound as to create four enormous lentil shapes virtually hiding the frame — is its paradigm.

With the completion of the "Trees," my sense of Winsor's work having keyed into the scheme of events diverges from her sense of keying into them. This requires some explanation. From 1973 on, it seems to me that Winsor has been engaged in a kind of revival paralleled in the work of contemporary sculptors of considerable reputation.

From my point of view, the artists who resist acknowledging that their work embodies issues, or ideas, or arguments, whatever else it surely does embody, these artists commit themselves to slower developments. Mind you, I have often noted that today's artistic consciousness at the moment of making an art that is deeply felt rejects the primacy of theory typical of high formalist abstraction. By contrast, artistic ratification is secured by the experimental piece, the work that bespeaks trial and error. Today, theory, if honored, is done so in the breach; but honor in the breach is also a theory of art.

It strikes me that something like this is presently affecting Winsor's work. The laminated and gouged plywoods of 1973 invoke the layerings and hammering processes of 1970. *Paul Walter's Piece* (1974), skeining copper wire around bundles of creosoted wood, recalls the *Solid Lattice* of 1970, as much as the hemp *Chunk Piece* of the same year. Just as Brancusi might be said to have tested acuity of sensibility in transferring a work from wood to stone to bronze, so too does it seem that Winsor's twinings of *#2 Copper* (1976) deal with such a parallel when compared to the *#2 Rope* (1976).

Similarly, the elegant wooden cubes of 1976 neatly invoke the incremental procedures and exhaustive additive processes of Winsor's early more expressively extroverted phrase. In each of the recent cubes, the serialized processes are self-evident, forming either solids and internal layers or regular interchanges layer to layer in their inevitably, somewhat dogmatic scholarly presentations.

In a certain sense this regularization of nice method is echoed in the recent layered steppings of Jackie Ferrara's curved pyramidal mounds. Only she, as Winsor notes, is involved in a "scale model sensibility." Ferrara's work assumes the character of architecture, whereas hers are literally *there* in the one, actual human scale.

Winsor's procedure is equally traceable to the organic character of any real artist's career: themes thread in and out to be revived and discarded as need be, or to exist in a multiple range of possibilities whereby a work in metal, say, alters our awareness of what is important in, say, one made of hemp. This conjures a method of shuttling comparison within a long-term thematic evolution.

What troubles me in the "organic argument," today especially, is the tacit assumption that artistic production is coeval with or built into a career, a long-term evolution. This may be so, but to me it is not necessarily so, at least not at this moment. Contrary to the "long career" premise — and Winsor is only in her mid-thirties, just entering mid-career — can we not at least entertain the notion that art may be as well in a "one-shot" event from which there is no "fallout," no sequence, no suite of things, no "career"?

The advent of the conceptual movement as the basis for urgent arts activity and its subsequent rapid dissolution into ever more theatrical and body referential modes of expression from which there is scant tangible "fallout" — no paintings, no sculpture — sufficiently justifies my perhaps passing attraction to the above possibility while indicating the continuous pressing intensity of our changing ideological climate.

Actually, in fairness it should be noted that Winsor had paid heed quite early on to non-objectified performance work in her *Up and/or Down Piece* (1971), a piece invoking "the deadweight of rope and the personality that rope takes on." (Liza Bear, *Avalanche,* 4, Spring 1972, p. 11.) In this theatrical work a male performer slowly pulled a vast amount of rope provided him by a female performer through an aperture in the floor, one story to another. This task completed, he lowered the rope through the aperture onto the body of the now recumbent female, at length burying her in its coils. The sexual nature of the piece, however ambiguous, was inescapable.

Thus, our moment, is marked by an essential divergence of attitude: one answering an artist's need to carry on the transcending logic of processes and substances leading to a continued production of objectified art; the other, perhaps primarily a critical attitude, being drawn by an alternative ironic range of expression.

# MICHAEL HURSON:
# A FABRIC OF AFFINITIES

Scarcely older than my students, I met my first class in the Ryerson auditorium of the Art Institute of Chicago. It was the fall of 1961. The memory is still fresh — an academic rite of passage. In that survey of 19th-century art were several young artists who would continue the history of The Chicago School, and through this local attachment, widen the issues of contemporary American art. The most singular, perhaps, was a quiet plain boy, Michael Hurson. As I watched him grow, he reshaped this shyness into an ironic ambiguity, into art that now seems a real alternative to the prevailing formalist hegemony or, at very least, to promote access to unexpected critical issues.

There was some priming. While a graduate student at the University of Chicago, I boarded with Thomas Lyman — the medieval art historian — and Mollie Michala — the painter — "Mr. and Mrs. Young Chicago" with at length six kids (they now live in Atlanta). They were enthusiastic about Hurson whom they had met shortly before at the Oxbow Art School in Saugatuck, Michigan, and who, by his funny manner, had become a friend of Burr Tillstrom (the originator of Kukla, Fran, and Ollie), then also summering at Oxbow. This connection would have considerable effect within Hurson's work, since for several years he earned his way assisting Tillstrom in the peregrinations of the Kuklapolitans. Much of the jokey private imagery which once seemed so peculiar in Hurson's work (when measured against Jansenist formalism) can be traced to this source — for example, the suite of blocky furniture, a take-off on the ranch style house in which he then lived, in the *Ballet of the Left-Handed Piano.*

Stories abounded, earning Hurson a grudged local celebrity. He had already appeared in the 1961 Annual. Through this exhibition, the artist met Henry Geldzahler, then a lion of Pop, recently appointed to the staff of the Metropolitan Museum of Art, and on tour looking at art across the country. Struck by Hurson's 1961 painting, he contacted the artist, then a sophomore at the School of the Art Institute of Chicago, and purchased a drawing from his sketch book. Hurson was just a kid without standing in the rigid hierarchy of the Chicago scene. Geldzahler's modest purchase added notoriety to the growing awareness — at least in art school — of Hurson's off-the-wall sensibility.

In my own way, I was also seen as an outsider, having come to the University of Chicago from New York City, interfacing my studies in the heartland with annual fellowships that took me to Paris, and, for a while, art dealing as Iris Clert's American assistant (Yves Klein, Takis, Tinguely, etc.) to flesh out the meager international fellowships. In the meanwhile, the School of the Art Institute had set up a new program: the advanced students' final year of painting or sculpture could be spent working alone with a single instructor. It was assumed that a painter would choose to work with a painter, a sculptor with a sculptor. Hurson asked to work with me, the *ausländer* art historian, a choice that rankled in a faculty that on the one hand stressed a retrograde Abstract Expressionism — Tenth Street on Michigan Avenue — and on the other, the ethical pretentions of Bauhaus refugees come to Chicago to begin again in the shadow of big Bauhaus names — Mies van der Rohe, Laszlo Moholy-Nagy, and the like. Both these factions, it went without saying, were distressed by one another; the Bauhaus group — headed by Paul Wieghardt, a respectable painter — feeling their ethos violated by the academized Abstract Expressionists — led by Charles Steegman, "Visiting Instructor" that year — and vice versa. Both were appalled — and thus momentarily united — by the ironies of the nascent Pop movement whose values were then being absorbed by a new generation of art school students.

What did Hurson do that senior year? His major effort was a suite of Barcelona chairs made out of schlock chicken wire and papier-mâché, somewhat akin to the early work of Claes Oldenburg. Since I was "just an art historian" and green, I was not asked to sit on Hurson's "Graduate Critique Committee" even though I was his "advisor." The popular imagery of furniture was anathema to the aniconic pretensions of the die-hard expressionists, and worse — that the exquisite chair of 1929, epitomizing so many of the ambitions of the Bauhaus enterprises, should have been guyed (or so it was thought) in the work of this suburban boy was felt to be an affront to lives shattered by the advent of Nazism. Of course that now lost work was none of these things. Hurson simply loved the Barcelona chair — it and other chromium designs have long functioned as preferred imagery for him — although he is not insensitive to the fact that the Mies look (consecrated by The Museum of Modern Art and Knoll International) was to emerge in the burgeoning steel and glass of the '50s and '60s as the essential style of corporate identity.

Among my papers I find a carbon of an angry letter to the committee that had diverted attention from Hurson's work to the broader and ambiguous problem of what was then being called "New Realism": ". . . should the Critique Committee force Michael Hurson into more conventional and pedantic channels they would not only be harmful to [Hurson's] artistic development, they would also reflect badly on the School of The Art Institute of Chicago, an institution traditionally above petty insularity." What fawning cant. It wasn't. They weren't. Hurson was put on probation, but with the possibility of an eleventh hour reprieve: a new group of senior projects were to be made based on works in the collection of The Art Institute. I remember when Hurson showed me a dumbly beautiful self-portrait based on the Fantin-Latour portrait of Manet. He would be redeemed. He had painted himself, replacing Manet's natty and glistening top hat with a folded paper hat, a child's party memory — a fool's cap — that since then still looms large in his iconography. Its connection to the costumes of Picasso's *Saltimbanques* is here maintained, and across these acrobats back through Daumier's theatre scenes of the *commedia dell'arte* to Watteau's portrait of *Gilles*.

Last year, after a stint in the Army, working as a draftsman in Maryland and Thailand, and a checkered sequence of addresses — New York, a return to Chicago — Hurson once more took up quarters in New York, a tiny, sky-lit studio space in a brownstone on East 10th Street — the type of building, classic to New York life, beloved of Henry James. Here was Michael again, a bit more scruffy and bunny-like, whose work continued to elaborate the complex imagery of fundamental social exchanges in generalizing, non-illustrative ways. In this theme, Hurson's work seems to key into facets of the broad range of meta-sociologies differently inflected in such work as Scott Burton's Behavioral Tableaux or Joel Shapiro's monumental doll-house furniture. Hurson's work, however, is not party to the crypto-mannerist and archly intellectual rejections of High Abstraction through conceptual theatrical exercises such as one finds in, say, Lynda Benglis or Robert Morris. It is also free of these accentuations as they inform many of the more photographic and conceptual manifestations rooted in the abstract principles of the later '60s, although he is among the first to have used photography in just this way — for example, the eight-part paper hat photographs dating to 1962. In short, Hurson's work is not critical stimulus/response art; rather it deals with a constant imagery that at its core is curiously uninflected and behaviorally neutral, white bread — WASP, spearmint, if you see what I mean — New Yorkers would think of it as Midwestern, down home. For the first time, Hurson's work which hitherto seemed will-o'-the wisp and arbitrary, is keying into modernist history; that's when the art may become important because it can be accounted for critically. Of course, such changes are in me; my stress on theory can now gather in an oeuvre which has always been organically whole, in touch with itself, and well able to survive according to its own light, with or without any critical prejudices.

What is equally clear is that the intentionist argument (art as a suite of conscious critical strategies) is not really opposed to sensibility art (at least not in the way it had been in the '60s through the early '70s). Indeed, neither manifestation is possible unless it grows from an organic center. In the end intentionist art becomes sensibility art — and is subsumed into the succession of historical styles — to which new intentions are attributed anyway as the work recedes further and further back from the present. Ironically, the real task of art history is to reconstruct the lost sense of intention pertinent to any given work.

Hurson was completing the last of his scrupulously made boxes, curious anonymous rooms made of balsa wood, this one based on the skylight structure of the very studio in which he lived. "Did you notice that the skylight shape is something like a paper hat?" This was the most recent of the rooms in a succession of constructions first shown at the Museum of Contemporary Art in Chicago (1973), and subsequently as one of the Projects series at The Museum of Modern Art (1975).

The Projects show further revealed the cryptic and biographical interlace of Hurson's work, culminating in his most elaborate and thwarting box, *Thurman Buzzard's Apartment* (Plate 49). I reviewed these constructions, trying to introduce the spectator to some of Hurson's elaborate built-in references. To pick up on the references in no way "explains" their fascination or effectiveness, just as knowing who Marie Taglioni was does not "explain" in any great degree the Joseph Cornell box in which references to the celebrated dancer are made. This kind of information is fascinating because it is embodied in art and suggests a direction for appreciating it; but it doesn't explain the art:

"The current works are miniature crafted sections of rooms, walls, window frames, and stairwells, although I still regard Hurson as a painter and a draftsman. Some balsa rooms are devoid of passing incident, while others are replete with bicycles, television sets, sofas, and evocative trappings. The most ambitious place, *Thurman Buzzard's Apartment,* addresses the psychic and physical space of an imaginary personage, an avatar of the artist, part fantasy, cryptogram and free association. Loosely speaking, Hurson equals Thurman and Buzzard equals Burr Tillstrom, a reference to Hurson's professional association with the celebrated puppeteer; the two r's in Burr equal the two z's in Buzzard, and so on.

". . . Implicit to the problem of miniaturization and the psychic paths that led Hurson to it, is a mind-set derived from Marcel Duchamp. Remember that Duchamp made a tiny replica of his *Nude Descending a Staircase* in 1918 for Carrie Stettheimer's dollhouse. And surely the Thorne Miniature Rooms at the Art Institute of Chicago provide an unexpected regional model for Hurson's constructions.

". . . What intrigues me in Hurson's constructions and his paintings is the interplay between an overt ordinariness of imagery and a covert system of logic. One without the other is merely netural; together in Hurson's work, these impulses produce among the most intriguing art I know being done today — even more striking for being couched in such plain terms." (*Artforum,* December 1974, p. 78.)

The review brings to mind that Hurson is, above all else, a painter, and like so much painting today his is realized in activities still assimilable to sculpture, or photography, or theatre, or even commercial art. In the early '60s, Hurson's paintings were marked by an expressionist purposefulness linked to sensibility, translations of the midwestern landscape — oil painting muddied by charcoal drawing, typical, say, of the Oxbow work. Still, certain personae began to appear at this time, anonymous figures, chunky furniture, or more autobiographically, the artist revealed as a paper hat or a polka dot tie. I appear as a mortar board — regional and attenuated versions of Duchamp's *Cemetery of Liveries and Uniforms* (1915), still sensible in a sociological identification through sartorial accoutrement.

This student period was followed by his military service, travels in Europe after his discharge, and still later, his first long sojourn in New York. This period (ca. 1964-66) is marked by little actual productivity, a certain dysfunction, a sense of dislocation, and the death of his father, "the person I most wanted to prove myself to." By 1967, a body of painting based on television imagery and associations emerges. Hurson had already been assisting Tillstrom, touring throughout the country — the Hollywood Palace, the Johnny Carson Show, then originating in New York. Hurson's preferred format became a long horizontal frieze, interspersed with benign Joes stretched out watching the tube. "I couldn't understand the notion of good composition. I had to figure out a way to make compositions that I could change easily so that I could correct them easily and figure out good composition that way; so I shellacked the canvas. Working on it with charcoal I can easily correct them like a blackboard."

The desire to shortcut and circumvent standardized notions of good composition led Hurson to make silkscreens of new images so that he could quickly squeegee them onto the canvas. "The subjects never changed; the composition did." Surely Warhol's and Rauschenberg's expressionist use of the silkscreen as a constant imagistic device is paralleled by this procedure in Hurson. Between 1969 and 1971, Dancing Eyeglasses and Pencils appear. The Pencil is the director marking locations where the actor should stand on narrow platforms; the Eyeglasses are the performers. Between 1971 and '72 the Palm Springs Swimming Pool emerges — four large panels depicting the image in 12 variations. The figure lying by the pool is Burr Tillstrom. Still executed through screens, these works invent visual parallel effects — locateable to television lininess — not dissimilar to certain hatchings seen in more recent painting by Johns.

Hurson's paintings to this time locate figures in physical settings that literally are the canvas grounds. By contrast, the balsa rooms emerging in '72 are free of personages although, as we have seen, they are conceived of as places in which avatars and personae live. These psychological surrogates led to a more recent focus on generalizing portraits such as the 1974 *Portrait of Edward and Otto Pfaff* (Plate 50). These portraits, like the other work squeegeed through screens, are a depersonalization — neither expressive nor illusionistic.

In the Pfaff portrait, "closest to Burr's ideas," two puppet-like characters, Otto and Edward, are distracted by some event. The full sequence leading to this momentary lapse of attention is absent. Only two frames are depicted — chosen from an elaborate now-abandoned jokebook-cum-storyboard (a TV and movie method) that dates from Hurson's student days. Edward is probably Edward Albee, whose early plays established short, prototypic expositions of social exchange. Otto, by simple letter substitutions, is the puppet Ollie, although physically he favors Kukla. Pfaff, the perfect name — those double l's and f's recalling Duchamp's attraction to the double r for his alter-ego Rrose Selavy, because of his delight with the double l's in Harold Lloyd's name — appeared on a sewing machine company's sign directly across the river from one of Hurson's Chicago studios.

The theme of absent personae continues from the Pfaff portrait to

*Thurman Buzzard's Apartment,* and informs a set of portraits currently being executed of the artist's friends as anonymous puppet-like figures. Hurson's system of reference, while it explains nothing of the art, still indicates much of a method through indirection that parallels through thought an expressionist system of mark, alteration, correction, and remarking. Hurson works out of this painterly methodology, throwing into its mechanism these memories and associations much as he might toss in a little more paint, a little more charcoal. In this sense, Hurson's fabric of affinities functions as one more tone, one more color on the painter's palette.

Grateful acknowledgment is extended to the following for permission to use their photographic material:
Mary Boone Gallery, Leo Castelli Gallery, Paula Cooper Gallery, Ronald Feldman Gallery, The Alberto Giacometti Foundation, The Hamilton Gallery, Gerald Hayes, Metro Pictures, Max Protetch Gallery, Sperone Westwater Fischer Inc., Ileana Sonnabend Gallery, Edward Thorp Gallery, John Weber Gallery, The Whitney Museum of American Art (New York).

## TITLES IN PRINT

Nanni Cagnone, *What's Hecuba to Him or He to Hecuba?* (poetry)
Richard Kostelanetz (ed.), *Essaying Essays. Alternative Forms of Exposition.* (literature)
Mario Diacono, Vito Acconci. *Dal testo-azione al corpo come testo.* (art criticism)
James Reineking, *Logical Space.* (art)
Giovanna Sandri, *From K to S. Ark of the Asymmetric. / Da K a S. Dimore dell'Asimmetrico* (poetry)
Tomaso Kemeny, *The Hired Killer's Glove. / Il guanto del sicario.* (poetry)
Rubina Giorgi, *Figure di Nessuno.* (philosophy)
Robert Pincus-Witten, *Postminimalism. American Art of the Decade.* (art criticism)
Giuseppe Risso, *Cospaia. The Practice of Essence.* (poetry)
Martino Oberto (OM), *Anaphilosophia. The Art of Language.* (philosophy)
Annina Nosei Weber (ed.), *Discussion. The Aesthetic Logomachy.* (art)
Abraham Lincoln Gillespie, *The Syntactic Revolution.* (literature)
Rosemary Mayer, *Pontormo's Diary.* (art)
Robert Pincus-Witten, *Entries (Maximalism). Art at the Turn of the Decade.* (art criticism)
Jeremy Gilbert-Rolfe, *Immanence and Contradiction. Recent Essays on the Artistic Device.* (art criticism)

## FORTHCOMING TITLES

Luigi Ballerini, *Che figurato muore.* (poetry)
Fredi Chiappelli (ed.), *Western Jerusalem. University of California Studies on Tasso.* (literary criticism)
Thomas J. Harrison (ed.), *The Favorite Malice. Ontology and Reference in Contemporary Italian Poetry.* (poetry and philosophy)
Angelo Lumelli (ed.), *Athenaeum.* (German literature)
Claudio Olivieri, *The Blind Predicate.* (art)
Richard Plunz, *Fifteen Families in Akcaalan. Notation toward an Anthropology for Building.* (architecture)